540

Chemistry Calculations Explained

Studymates

British History 1870–1918
Warfare 1792–1918
Hitler and Nazi Germany (3rd Edition)
English Reformation
European History 1870–1918
Genetics (2nd edition)
Lenin, Stalin and Communist Russia
Organic Chemistry
Chemistry: As Chemistry Explained
Chemistry: Chemistry Calculations Explained
The New Science Teacher's Handbook
Mathematics for Adults
Calculus
Understanding Forces
Algebra: Basic Algebra Explained
Plant Physiology
Poems to Live By
Shakespeare
Poetry
Better English
Better French
Better German
Better Spanish
Social Anthropology
Statistics for Social Science
Practical Drama and Theatre Arts
The War Poets 1914–18
The Academic Essay
Your Master's Thesis
Your PhD Thesis

Chemistry Calculations Explained

A Ninan PhD CChem MRSC

www.studymates.co.uk

© 2005 by A Ninan

ISBN 10 1-84 285-072-5
ISBN 13 978 1-84 285-072-5

First published in 2005 by Studymates Limited.
PO Box 225, Abergele, LL18 9AY, United Kingdom.

Website: http://www.studymates.co.uk

Typeset by Domex e-Data Pvt. Ltd.
Printed and bound in Great Britain by Baskerville Press

Contents

Preface

My experience as a teacher and examiner in Chemistry for over twenty years has shown me that a lot of students struggle with the mathematical aspects of chemistry and this is clearly manifested in the papers I mark every summer. I wanted to write a book that both simplified and demystified this aspect of Chemistry. This book is ideal for A level students and for those in the early years of degree level chemistry.

Each chapter begins with the 'one-minute overview' followed by important definitions and the relevant equations required for the calculations. Carefully thought out sample calculations are provided, laid out in increasing order of difficulty. This will help students build confidence and understanding. Finally a set of practice questions is provided for students to try out the concepts that they have learnt with answers at the end of the book.

I sincerely hope that this book will be successful in clarifying what may seem to many students an abstruse side of Chemistry. I welcome all feedback from students, their teachers and any other readers.

I would like to thanks Dr Golda Ninan for reading the book and giving valuable suggestions.

Dr A. Ninan

E-mail: *a.ninan@studymates.co.uk*

The mole concept and stoichiometry

One-minute overview

The *mole concept* provides a useful framework to calculate quantities of substances involved in reactions. A *chemical equation* is a starting point for determining mole ratios and masses. This chapter starts with an outline of important definitions to act as a reference point for the calculations which follow.

The points covered include:
- relative atomic and molecular masses
- the mole and Avogadro's constant
- the molar volume of gases
- stoichiometric equations and masses of reactants and products
- reactions in solution.

1.1 Basic definitions and relationships

Relative atomic mass and relative molecular mass have no units.

The **relative atomic mass**, A_R, of an element is the mass of one atom of it relative to the mass of one atom of ^{12}C, which has a mass of exactly 12 units (12 u) on the **atomic mass unit scale**. Since an element can be a mixture of two or more isotopes, the relative atomic mass of the element is the weighted mean of the isotopic masses of all its isotopes (see Example 1.1).

The **relative molecular mass**, M_R, of a substance is the sum of the relative atomic masses of all the atoms in a molecule of that substance. Since ionic compounds contain no molecules but ions, the term **relative formula mass** is often used instead of relative molecular mass.

The **mole** (mol) is that quantity of a substance which contains as many particles (atoms, molecules or ions) as there are atoms in exactly 12 g of ^{12}C.

Avogadro's constant (or **Avogadro's number**), L, is the number of atoms present in 12 g of ^{12}C. The value of Avogadro's constant is 6.022×10^{23} mol^{-1}.

The **molar mass, M,** of a substance is the mass of 1 mole of that substance in grams.

The unit for molar mass is g mol^{-1}.

$$M = M_R \text{ g mol}^{-1}$$

The **relationship between mass (m) and molar mass (M)** is given by

$$\text{number of moles } (n) = \frac{\text{mass}}{\text{molar mass}}$$

$$n = \frac{m}{M}$$

The **molar volume** is the volume occupied by 1 mole of a gas and its value is 22.4 dm^3 for any gas at standard temperature and pressure (stp). Standard temperature is 0°C (273 K) and standard pressure 1 atmosphere (1 atm)

1 atm = 1.01 × 10^5 Pascal (Pa) = 1.01 × 10^5 N m^{-2} (See Section 2.1 for a discussion of the units of pressure).

At room temperature (20°C or 293 K) and 1 atm pressure (rtp) the value of the molar volume is approximately 24.0 dm^3.

Relationship between moles and stoichiometric equations

A stoichiometric equation (balanced chemical equation) provides the number of molecules that react and that are formed. This information can be used to deduce the number of moles of reactants and products involved in a chemical reaction. For example, consider the reaction between aluminium and hydrochloric acid. The balanced equation for the reaction is

$$2Al(s) + 6HCl(aq) \rightarrow 2AlCl_3(aq) + 3H_2(g)$$

2 moles of Al react with 6 moles of HCl to give 2 moles of AlCl$_3$ and 3 moles of H$_2$. Using this information, the mass ratio for the reaction can be calculated. (Relative atomic

mass, A_R: H = 1.0, Al = 27.0, Cl = 35.5; relative molecular mass, M_R: H_2 = 2.0, HCl = 36.5, $AlCl_3$ = 133.5)

$$2Al(s) \ + \ 6HCl\,(aq) \ \rightarrow \ 2AlCl_3\,(aq) \ + \ 3H_2\,(g)$$

2 mol	6 mol	2 mol	3 mol
2×27.0 g	6×36.5 g	2×133.5 g	3×2.0 g

That is, 54.0 g of Al reacts with 219.0 g of HCl to give 267.0 g of $AlCl_3$ and 6.0 g of H_2.

Percentage yield

When a substance is prepared by chemical reactions, the yield of the substance is usually lower than the calculated (theoretical) yield. The **percentage yield** is a useful way of expressing this:

$$\text{percentage yield} = \frac{\text{experimental yield}}{\text{theoretical yield}} \times 100$$

Percentage purity

It is sometimes useful to express the **purity** of a substance in terms of its proportion in an impure sample:

$$\text{percentage purity} = \frac{\text{mass of the pure substance}}{\text{mass of the impure sample}} \times 100$$

The **concentration** of a solution is often expressed as the number of moles of a substance present in one cubic decimetre of solution. For example, if 2.0 moles of sulphuric acid is present in 1 dm^3 (litre) of solution, its concentration is 2.0 mol dm^{-3}.

> A molar solution is one which contains 1 mole of solute in 1 dm^3 of solution.

The **molarity (molar concentration)** of a solution is the number of moles of solute present in 1 dm^3 of solution. One of the two following equations can be used to calculate the molarity of a solution:

$$\text{molarity} = \frac{\text{concentration in g } dm^{-3}}{\text{molar mass}}$$

$$\text{molarity} = \frac{\text{no. of moles of the solute}}{\text{volume of solution in } dm^3}$$

> The unit for the molarity (molar concentration) of a solution is mol dm^{-3}.

1.2 Calculations involving mass, the mole and Avogadro's constant

The data given below are used in the calculations in the remainder of this chapter.

Element	Relative atomic mass	Element	Relative atomic mass
H	1.0	S	32.1
He	4.0	Cl	35.5
Li	6.9	K	39.1
C	12.0	Ar	40.0
N	14.0	Ca	40.1
O	16.0	Fe	55.9
F	19.0	Cu	63.6
Ne	20.2	Br	79.9
Na	23.0	I	126.9
Mg	24.3		

The molar volume of a gas at room temperature (293 K) and pressure (1 atm) = 24.0 dm^3

The molar volume of a gas at standard temperature (273 K) and pressure (1 atm) = 22.4 dm^3

Avogadro's constant = 6.022×10^{23} mol^{-1}

Example 1.1 Calculating relative atomic mass

If a sample of copper contains 69.1% of ^{63}Cu of isotopic mass 62.930 u and 30.9% of ^{65}Cu of isotopic mass 64.928 u, calculate the relative atomic mass of copper.

Solution

Relative atomic mass of copper

$$= \frac{(69.1 \times 62.930) + (30.9 \times 64.928)}{100}$$

$$= 63.547$$

Example 1.2 Calculating relative molecular mass (formula mass)

Calculate the relative molecular mass of these substances.

(a) Cl_2,   (b) KBr,   (c) $CaSO_4 \cdot 2H_2O$

Solution

Relative molecular mass, M_R:

(a) Cl_2 $= 2 \times 35.5$ $= 71.0$

(b) KBr $= 39.1 + 79.9 = 119.0$

(c) $CaSO_4 \cdot 2H_2O$
$$= 40.1 + 32.1 + (4 \times 16.0) + [2 \times (2 \times 1.0 + 16.0)]$$
$$= 172.2$$

Example 1.3 Converting moles to mass

Calculate the mass, in grams, of 0.05 mol of $NaHCO_3$.

Solution

$$M\,(NaHCO_3) = 84.0 \text{ g mol}^{-1}$$

$$\text{Mass} = \text{no. of moles} \times \text{molar mass}$$

$$= 0.05 \text{ mol} \times 84.0 \text{ g mol}^{-1}$$

$$= 4.2 \text{ g}$$

Units can be modified by multiplication, division etc. So, 'mol' will cancel out 'mol^{-1}'. Thus mol $\times$ g mol^{-1} = g.

Example 1.4 Converting number of molecules to mass

Calculate the mass, in grams, of 3.0×10^{22} molecules of Br_2

Solution

1 mol of Br_2 contains 6.022×10^{23} molecules

$$\text{Moles of } Br_2 = \frac{3.0 \times 10^{22}}{6.022 \times 10^{23}}$$

$$= 4.98 \times 10^{-2} \text{ mol}$$

$$M\,(\text{Br}_2) = 159.8 \text{ g mol}^{-1}$$

$$\text{Mass of Br}_2 = 4.98 \times 10^{-2} \text{ mol} \times 159.8 \text{ g mol}^{-1}$$

$$= 7.96 \text{ g}$$

Example 1.5 Converting mass to moles

Calculate the number of moles in the following amounts of substances.

(a) 2.8 g of Fe,   (b) 100 g of NaOH

Solution

(a) $$M\,(\text{Fe}) = 55.9 \text{ g mol}^{-1}$$

$$\text{Moles of Fe} = \frac{2.8 \text{ g}}{55.9 \text{ g mol}^{-1}}$$

$$= 0.050 \text{ mol}$$

(b) $$M\,(\text{NaOH}) = 40.0 \text{ g mol}^{-1}$$

Note the modification of the units here.

$$\frac{g}{g\,\text{mol}^{-1}} = \frac{\cancel{g}}{\cancel{g}\,\text{mol}^{-1}}$$
$$= \frac{1}{\text{mol}^{-1}}$$
$$= \text{mol}$$

$$\text{Moles of NaOH} = \frac{100 \text{ g}}{40.0 \text{ g mol}^{-1}}$$

$$= 2.50 \text{ mol}$$

Example 1.6 Calculating the number of particles

Calculate the number of atoms, molecules or ions in the following substances.

(a) Atoms in 0.018 g of He

(b) Molecules in 0.020 mol of CO_2

(c) H^+ and Cl^- ions in a solution containing 3.65 g of HCl

Solution

(a) $$\text{Moles of He} = \frac{0.018 \text{ g}}{4.0 \text{ g mol}^{-1}} = 4.5 \times 10^{-3} \text{ mol}$$

1 mol of He contains 6.022×10^{23} atoms.

Number of atoms in 4.5×10^{-3} mol of He

$$= (4.5 \times 10^{-3} \text{ mol}) \times (6.022 \times 10^{23} \text{ mol}^{-1})$$

$$= 2.70 \times 10^{19}$$

(b) 1 mol of CO_2 contains 6.022×10^{23} molecules.

Number of molecules in 0.020 mol of CO_2

$$= (0.020 \text{ mol}) \times (6.022 \times 10^{23} \text{ mol}^{-1})$$

$$= 1.20 \times 10^{22}$$

(c) Molar mass of HCl $= 36.5$ g mol^{-1}

$$\text{Moles of HCl} = \frac{3.65 \text{ g}}{36.5 \text{ g mol}^{-1}} = 0.100 \text{ mol}$$

0.1 mole of HCl produces 0.1 mole of H^+ and 0.1 mole of Cl^- ions.

Number of H^+ ions $= (0.100 \text{ mol}) \times (6.022 \times 10^{23} \text{ mol}^{-1})$

$$= 6.02 \times 10^{22}$$

Number of Cl^- ions $= (0.100 \text{ mol}) \times (6.022 \times 10^{23} \text{ mol}^{-1})$

$$= 6.02 \times 10^{22}$$

1.3 Calculations involving the molar volume of gases

Example 1.7 Calculating moles from the volume of a gas

Calculate the number of moles of N_2 in 10.0 dm^3 of N_2 at 1 atm and 293 K.

Solution

24.0 dm^3 of N_2 at 1 atm and 293 K contain 1 mol of N_2.

Moles of N_2 in 10.0 dm^3 at the same temperature and pressure

$$= \frac{1 \text{ mol} \times 10.0 \text{ dm}^3}{24.0 \text{ dm}^3} = 0.417 \text{ mol}$$

Example 1.8 Calculating volumes of gases from the number of moles

Calculate the volumes of these gases at stp:

(a) 0.005 mol of Ar,   (b) 50.0 g of Cl_2

Solution

(a) Volume occupied by 1 mol of a gas
$$= 22.4 \text{ dm}^3 \text{ at stp}$$

Volume of 0.005 mol of Ar at stp
$$= (0.005 \text{ mol}) \times (22.4 \text{ dm}^3 \text{ mol}^{-1})$$

$$= 0.112 \text{ dm}^3$$

(b) Molar mass of $Cl_2 = 2 \times 35.5 = 71.0 \text{ g mol}^{-1}$

$$\text{Moles of } Cl_2 = \frac{50.0 \text{ g}}{71.0 \text{ g mol}^{-1}}$$

$$= 0.704 \text{ mol}$$

Volume of Cl_2 at stp
$$= (0.704 \text{ mol}) \times (22.4 \text{ dm}^3 \text{ mol}^{-1})$$

$$= 15.8 \text{ dm}^3$$

1.4 Calculations based on stoichiometric equations

Example 1.9 Calculating the number of moles of a product

Calculate the number of moles of hydrogen produced by the reaction of 5.6 g of lithium metal with water, assuming there is enough water for the completion of the reaction.

Solution

Step 1: Write a balanced equation for the reaction between Li and H_2O and work out the mole ratio.

$$2Li(s) + 2H_2O(l) \rightarrow 2LiOH(s) + H_2(g)$$

2 mol   1 mol

Step 2: Calculate the number of moles in 5.6 g of Li.

$$\text{Moles of Li} = \frac{5.6\ g}{6.9\ g\ mol^{-1}} = 0.81\ mol$$

Step 3: From the stoichiometric equation, it is clear that 1 mole of H_2 is produced from 2 moles of Li.

$$\text{Moles of } H_2 \text{ gas} = \frac{0.81\ mol}{2} = 0.405\ mol$$

Example 1.10 Calculating reacting masses

In the Thermite process, aluminium reduces iron(III) oxide to form iron and aluminium oxide.

$$2Al(s) + Fe_2O_3(s) \rightarrow Al_2O_3(s) + 2Fe(s)$$

Calculate the mass of aluminium required to reduce 100 g of iron(III) oxide.

Solution

$$2Al(s) + Fe_2O_3(s) \rightarrow Al_2O_3(s) + 2Fe(s)$$

2 mol   1 mol

2×27.0 g   159.8 g

159.8 g of Fe_2O_3 is reduced by 54.0 g of Al.

100 g of Fe_2O_3 is reduced by

$$\frac{54.0\ g \times 100\ g}{159.8\ g} = 33.8\ g \text{ of Al}$$

Example 1.11 Calculating masses and/or volume of products

Magnesium reacts with hydrochloric acid to give magnesium chloride and hydrogen. Magnesium chloride can in turn be crystallised out of this reaction mixture as $MgCl_2 \cdot 6H_2O(M_R = 203.3)$

(a) What mass of $MgCl_2 \cdot 6H_2O$ can be prepared starting from 2.20 g of magnesium metal and hydrochloric acid, assuming that there is enough HCl to react with all the metal?

(b) What volume of H_2 gas at room temperature (293 K) and pressure (1 atm) is produced at the same time?

Solution

(a) Moles of Mg reacted $= \dfrac{2.20 \text{ g}}{24.3 \text{ g mol}^{-1}} = 0.091 \text{ mol}$

$$Mg(s) + 2HCl(aq) \rightarrow MgCl_2(aq) + H_2(g)$$

1 mol $\qquad\qquad$ 1 mol $\qquad$ 1 mol

$$MgCl_2 + 6H_2O \rightarrow MgCl_2 \cdot 6H_2O$$

1 mol $\qquad\qquad$ 1 mol

Moles of $MgCl_2 \cdot 6H_2O$ formed
$$= 0.091 \text{ mol}$$

Molar mass of $MgCl_2 \cdot 6H_2O$
$$= 203.3 \text{ g mol}^{-1}$$

Mass of $MgCl_2 \cdot 6H_2O$ formed
$$= (0.091 \text{ mol}) \times (203.3 \text{ g mol}^{-1})$$

$$= 18.50 \text{ g}$$

(b) Moles of H_2 produced $=$ moles of Mg
$$= 0.091 \text{ mol}$$

Volume of H_2 at 293 K and 1 atm
$$= (0.091 \text{ mol}) \times (24.0 \text{ dm}^3 \text{ mol}^{-1})$$

$$= 2.18 \text{ dm}^3$$

1.5 Calculations involving reactions in solution

Example 1.12 Calculating molarity (molar concentration)

2 dm^3 of an acid solution contains 4.9 g of H_2SO_4. Calculate the concentration, in $mol \, dm^{-3}$, of:

(a) H_2SO_4, (b) H^+ ions and (c) SO_4^{2-} ions.

((M_R: H_2SO_4 = 98))

Solution

(a) The acid solution contains 4.9 g of H_2SO_4 in 2 dm^3 of solution

$$\text{Concentration of } H_2SO_4 = \frac{4.9 \text{ g}}{2 \text{ dm}^3} = 2.45 \text{ g dm}^{-3}$$

Molar concentration of H_2SO_4

$$= \frac{2.45 \text{ g dm}^{-3}}{98 \text{ g mol}^{-1}} = 0.025 \text{ mol dm}^{-3}$$

(b) One H_2SO_4 molecule dissociates to give two H^+ ions and one SO_4^{2-} ion.

$$\text{Concentration of } H^+ = 2 \times 0.025 \text{ mol dm}^{-3}$$

$$= 0.050 \text{ mol dm}^{-3}$$

(c) Concentration of SO_4^{2-} = 0.025 mol dm^{-3}

Example 1.13 Calculating the mass of a solute from concentration

Calculate the mass of $Ca(HCO_3)_2$ present in 500.0 cm^3 of 1.0×10^{-3} mol dm^{-3} solution. (M_R: $Ca(HCO_3)_2$ = 162.1)

Solution

Method 1: Molarity of the solution

$$= 1.0 \times 10^{-3} \text{ mol dm}^{-3}$$

That is, 1 dm^3 solution contains 1.0×10^{-3} mol of $Ca(HCO_3)_2$

500 cm^3 solution contains

$$= \frac{1.0 \times 10^{-3}}{2} = 5.0 \times 10^{-4} \text{ mol}$$

$$\text{Mass of } Ca(HCO_3)_2 = (5.0 \times 10^{-4} \text{ mol}) \times (162.1 \text{ g mol}^{-1})$$

$$= 0.081 \text{ g}$$

Method 2:

$$\text{Molarity} = \frac{\text{moles of solute}}{\text{volume of solution in dm}^3}$$

Moles of $Ca(HCO_3)_2$
$$= \text{molarity} \times \text{volume of solution in dm}^3$$
$$= (1.0 \times 10^{-3} \text{ mol dm}^{-3}) \times (0.500 \text{ dm}^3)$$
$$= 5.0 \times 10^{-4} \text{ mol}$$

Mass of $Ca(HCO_3)_2$
$$= 0.081 \text{ g (as calculated above)}$$

Example 1.14 Calculating the concentration of a solution

A concentrated acid contains 12.0 mol dm^{-3} HCl. 200 cm^3 of this acid is diluted to 5 dm^3 with water. What is the concentration of HCl in the resultant solution?

Solution

Moles of HCl in the concentrated acid
$$= 12.0 \text{ mol dm}^{-3}$$

Moles of HCl in 0.2 dm^3 (200 cm^3) of this acid
$$= (12 \text{ mol dm}^{-3}) \times (0.2 \text{ dm}^3)$$
$$= 2.4 \text{ mol.}$$

So, 5 dm^3 of the diluted solution contains 2.4 mol of HCl.

$$\text{Concentration of HCl} = \frac{2.4 \text{ mol}}{5 \text{ dm}^3} = 0.48 \text{ mol dm}^{-3}$$

Example 1.15 Calculating the volume of a solution needed for a reaction

A solution contains 2.0 g of HNO_3 in 200 cm^3 of solution. Calculate:

(a) the concentration of the solution in mol dm^{-3}

(b) the volume of 0.11 mol dm^{-3} Ba(OH)$_2$ required to neutralise 50.0 cm^3 HNO$_3$. (M_R: HNO$_3$ = 63)

Solution

(a) 200 cm^3 (0.2 dm^3) of solution contains 2.0 g of HNO$_3$.

Moles of HNO$_3$ in 0.2 dm^3 solution

$$= \frac{2.0 \text{ g}}{63 \text{ g mol}^{-1}} = 0.0317 \text{ mol}$$

Molar concentration of HNO$_3$
$$= 0.0317 \times 5 = 0.159 \text{ mol dm}^{-3}$$

(b) Moles of HNO$_3$ in 50 cm^3 of the solution

$$= \frac{0.159 \text{ mol} \times 50 \text{ cm}^3}{1000 \text{ cm}^3}$$

$$= 0.00795 \text{ mol}$$

2HNO$_3$(aq) + Ba(OH)$_2$(aq) $\rightarrow$ Ba(NO$_3$)$_2$(aq) + 2H$_2$O(1)
2 mol   1 mol

Mole ratio, HNO$_3$: Ba(OH)$_2$ = 2 :1

Moles of Ba(OH)$_2$ required to neutralise 0.00795 mol of HNO$_3$

$$= \frac{0.00795 \text{ mol}}{2} = 0.00398 \text{ mol}$$

Now, the molarity of the Ba(OH)$_2$ solution
$$= 0.11 \text{ mol dm}^{-3}$$

That is, 0.11 mol of Ba(OH)$_2$ is present in 1 dm^3 (1000 cm^3) of the solution.

Therefore, 0.00398 mol of Ba(OH)$_2$ is present in
$$\frac{1000 \text{ cm}^3 \times 0.00398 \text{ mol}}{0.11 \text{ mol}} = 36.2 \text{ cm}^3$$

36.2 cm^3 Ba(OH)$_2$ solution is needed to neutralise 50.0 cm^3 HNO$_3$

Example 1.16 Calculations involving the volume and concentration of solutions

25 cm³ of 0.12 mol dm⁻³ $NaHCO_3$ solution is neutralised by 40.4 cm³ of H_2SO_4 solution. Calculate the molar concentration of the sulphuric acid solution.

Solution

Concentration of $NaHCO_3$ solution

$$= 0.12 \text{ mol dm}^{-3}$$

Moles of $NaHCO_3$ in 25 cm³ of the solution

$$= \frac{0.12 \text{ mol} \times 25 \text{ cm}^3}{1000 \text{ cm}^3} = 0.003 \text{ mol}$$

$$2NaHCO_3(aq) + H_2SO_4(aq) \rightarrow Na_2SO_4(aq) + 2H_2O(l) + 2CO_2(g)$$

2 mol   1 mol

Moles of H_2SO_4 to neutralise 0.003 mol of $NaHCO_3$

$$= \frac{0.003}{2} = 0.0015 \text{ mol}.$$

40.4 cm³ of the acid solution contains 0.0015 mol of H_2SO_4

1000 cm³ of the solution contains

$$\frac{0.0015 \text{ mol} \times 1000 \text{ cm}^3}{45.0 \text{ cm}^3} = 0.033 \text{ mol of } H_2SO_4$$

Molar concentration of H_2SO_4

$$= 0.033 \text{ mol dm}^{-3}$$

Example 1.17

A solution containing Fe^{2+} ions is prepared by dissolving 2.5 g of iron(II) sulphate crystals ($FeSO_4 \cdot 7H_2O$) in a mixture of water and dilute sulphuric acid to produce 200 cm³ of solution. 25 cm³ of this solution required 16.5 cm³ of

$KMnO_4$ solution for titration. The equation for the reaction between Fe^{2+} and MnO_4^- ions in the presence of acid is

$$5Fe^{2+}(aq) + MnO_4^-(aq) + 8H^+(aq) \rightarrow 5Fe^{3+}(aq)$$
$$+ Mn^{2+}(aq) + 4H_2O(l)$$

(M_R: $FeSO_4 \cdot 7H_2O = 298$)

Calculate:

(a) the concentration of Fe^{2+} ions in mol dm^{-3},

(b) number of moles of Fe^{2+} ions in 25 cm^3 of the solution,

(c) the number of moles of MnO_4^- in 16.5 cm^3 of the solution,

(d) the concentration of $KMnO_4$ in mol dm^{-3}.

Solution

200 cm^3 (0.2 dm^3) of solution contains 2.5 g of $FeSO_4 \cdot 7H_2O$

Moles of $FeSO_4 \cdot 7H_2O$
$$= \frac{2.5\ g}{298\ g\ mol^{-1}} = 8.39 \times 10^{-3}\ mol$$

Concentration of $FeSO_4 \cdot 7H_2O$
$$= 8.39 \times 10^{-3} \times 5 = 0.042\ mol\ dm^{-3}$$

(a) Concentration of Fe^{2+}
$$= 0.042\ mol\ dm^{-3}$$

(b) Moles of Fe^{2+} in 25 cm^3 of solution
$$= \frac{0.042 \times 25}{1000} = 0.00105\ mol$$

$$5Fe^{2+}(aq) + MnO_4^-(aq) + 8H^+(aq) \rightarrow 5Fe^{3+}(aq)$$
$$+ Mn^{2+}(aq) + 4H_2O(l)$$
5 mol   1 mol

(c) Moles of MnO_4^- required to react with 0.00105 mol of

$$Fe^{2+} = \frac{0.00105}{5}$$

$$= 2.1 \times 10^{-4}\ mol$$

That is, 2.1×10^{-4} mol of MnO_4^- is present in 16.5 cm³ of the solution.

(d) 16.5 cm³ of solution contains 2.1×10^{-4} mol of MnO_4^- ions.

1000 cm³ of solution contains
$$\frac{2.1 \times 10^{-4} \times 1000}{16.5} = 0.0127 \text{ mol}$$

Concentration of $KMnO_4$ solution = 0.0127 mol dm⁻³

1.6 Practice questions

Use the data on page 4 to answer these questions.

1. A sample of magnesium contains 79.0% of ^{24}Mg of isotopic mass 23.986 u, 10.0% of ^{25}Mg of isotopic mass 24.985 u and 11.0% of ^{26}Mg of isotopic mass 25.985 u. Calculate the relative atomic mass of Mg.

2. Calculate the relative atomic mass of chlorine if it contains 75.75% of ^{35}Cl of isotopic mass 34.970 u and 24.25% of ^{37}Cl of isotopic mass 36.969 u.

3. Calculate the molar mass (formula mass) of the following substances, giving the unit in each case.

   (a) Sodium carbonate decahydrate, $Na_2CO_3 \cdot 10H_2O$

   (b) Glucose, $C_6H_{12}O_6$

   (c) 1-Bromobutane, C_4H_9Br

   (d) Argon, Ar

   (e) Sulphur trioxide, SO_3

   (f) Fluorine, F_2

   (g) Sulphate ion, SO_4^{2-}

   (h) Tetraamine copper(II) ion, $[Cu(NH_3)_4]^{2+}$

   (i) Potassium hexacyanoferrate(III), $K_3[Fe(CN)_6]$

4. Calculate the number of moles in

   (a) 5.0 g of H_2 gas

   (b) 4.8 dm^3 of Cl_2 gas at 20°C and 1 atm

   (c) 3.5×10^{22} atoms of Ne

   (d) 2.0 g of $CaCO_3$

5. Calculate the mass in grams of

   (a) 0.15 mol of NaCl

   (b) 0.10 mol of $CuSO_4 \cdot 5H_2O$

   (c) 0.02 mol of Br_2 gas

   (d) 3.011×10^{25} molecules of CO_2

6. Calculate the number of molecules in

   (a) 9.6 g of SO_2 gas

   (b) 9.2 g of ethanol, C_2H_5OH

7. What mass of calcium oxide can be obtained by the complete decomposition of 40.04 g of calcium carbonate?

$$CaCO_3(s) \rightarrow CaO(s) + CO_2(g)$$

8. Calculate the mass of copper(II) nitrate needed to obtain 2.0 g of copper(II) oxide. Copper(II) nitrate decomposes completely to give copper(II) oxide, nitrogen dioxide and oxygen, according to the equation

$$2Cu(NO_3)_2(s) \rightarrow 2CuO(s) + 4NO_2(g) + O_2(g)$$

9. Potassium carbonate reacts with hydrochloric acid according to the equation

$$K_2CO_3(s) + 2HCl(aq) \rightarrow 2KCl(aq) + H_2O(l) + CO_2(g)$$

6.91 g of K_2CO_3 solid is treated with excess hydrochloric acid. Calculate the volume of carbon dioxide produced at room temperature (293 K) and pressure (1 atm).

10. A 100 cm³ solution contains 6.1 g of $MgCl_2 \cdot 6H_2O$.

   (a) Calculate the total number of moles of ions in the solution, ignoring the ions formed by the slight dissociation of water.

   (b) What is the concentration of Mg^{2+} ions in mol dm⁻³ ? (Hint: $MgCl_2$ dissociates to give one Mg^{2+} ion and two Cl^- ions.)
   [M ($MgCl_2 \cdot 6H_2O$) = 203.3 g mol⁻¹]

11. What volume of 11.7 mol dm⁻³ HCl should be taken to prepare 2.0 dm³ of 0.1 mol dm⁻³ hydrochloric acid?

12. What volume of water should be added to 1 dm³ of 10.0 mol dm⁻³ H_2SO_4 to make its concentration 5.0 mol dm⁻³?

13. 4.03 g of magnesium oxide is treated with 100.0 cm³ of 0.5 mol dm⁻³ hydrochloric acid.

$$MgO(s) + 2HCl(aq) \rightarrow MgCl_2(aq) + H_2O(l)$$

Calculate

   (a) the number of moles in 4.03 g of MgO

   (b) the number of moles of HCl taken up

   (c) the number of moles of the excess reagent.

14. Calculate the volume of 0.2 mol dm⁻³ H_2SO_4 needed to neutralise 200 cm³ of 0.5 mol dm⁻³ KOH solution.

$$H_2SO_4(aq) + 2KOH(aq) \rightarrow K_2SO_4(aq) + 2H_2O(l)$$

15. 50.0 cm³ of a saturated solution of calcium hydroxide requires 75.0 cm³ of 0.02 mol dm⁻³ HCl for neutralisation.

$$Ca(OH)_2(aq) + 2HCl(aq) \rightarrow CaCl_2(aq) + 2H_2O(l)$$

Calculate:

   (a) the number of moles of HCl required for neutralisation

   (b) the number of moles of $Ca(OH)_2$ in 50 cm³ of solution

(c) the solubility of $Ca(OH)_2$ in mol dm^{-3}

(d) the solubility of $Ca(OH)_2$ in g dm^{-3}.

16. 8.0 g of an impure sample of sodium hydrogen carbonate requires 100.0 cm^3 of 0.40 mol dm^{-3} HCl for complete neutralisation. [M ($NaHCO_3$) = 84.0 g mol^{-1}]

$$NaHCO_3(s) + HCl(aq) \rightarrow NaCl(aq) + CO_2(g) + H_2O(l)$$

Assuming that the impurities do not react with the acid, calculate

(a) the number of moles of HCl taken

(b) the number of moles of $NaHCO_3$ reacted

(c) the mass of $NaHCO_3$ reacted

(d) the percentage purity of the sample of $NaHCO_3$.

2 Gases: pressure, volume and temperature relationships

One-minute overview

A gas has no definite shape and it fills the container which holds it. Gas molecules move randomly in all directions. The forces of attraction between the molecules in a gas are weaker than those in liquids and solids. A gas is compressible and its volume is pressure dependent. Raising the temperature of a gas can increase its volume or its pressure or both.

This chapter deals with important laws governing the pressure, volume and temperature of gases and includes calculations based on:

- Boyle's law
- Charles' law
- Avogadro's law
- Dalton's law of partial pressures
- the number of moles present in a gas
- the relative molecular mass of a gaseous substance.

2.1 Basic definitions and relationships

Boyle's law states that the volume (V) of a fixed mass of gas is inversely proportional to its pressure (p) at constant temperature. This is expressed mathematically as,

$$V \propto \frac{1}{p}$$

that is, $pV = k$ (where k is a constant)

Therefore,

$$p_1 V_1 = p_2 V_2$$

where p_1 and p_2 are the initial and final pressures, and V_1 and V_2 the initial and final volumes.

Charles' law states that at constant pressure, the volume (V) of a fixed mass of gas is directly proportional to its temperature (T) in Kelvin.

$$V \alpha\ T$$
$$\frac{V_1}{T_1} = \frac{V_2}{T_2}$$

Combining Boyle's and Charles' laws, the following **general gas equation** can be derived,

$$\frac{p_1 V_1}{T_1} = \frac{p_2 V_2}{T_2}$$

Avogadro's law states that at constant temperature and pressure, the volume (V) of a gas is directly proportional to the number of moles (n) of the gas,

$$V \alpha\ n$$

The important implication of Avogadro's law is that under the same conditions of temperature and pressure, equal volumes of *all* gases contain the same number of moles (molecules).

Dalton's law of partial pressures states that the total pressure of a mixture of gases is equal to the sum of partial pressures of the individual gases. The partial pressure of a gas is the pressure exerted by it, if it alone were present in the container at the same temperature.

The values of the partial pressures of each of the gases in a mixture can be calculated if you know the total pressure of the mixture and the **mole fractions**, x, of the components. If a mixture contains two or more gases, one of which is A, the mole fraction of that component, x_A, is calculated by

$$x_A = \frac{\text{number of moles of A}}{\text{total number of moles}} = \frac{n_A}{n_{total}}$$

The partial pressue of A is equal to its mole fraction times the total pressure,

$$p_A = x_A \times p_{total}$$

The **ideal gas law** is $pV = nRT$, where n is the number of moles of the gas and R is the **universal gas constant**. A value for R can be calculated knowing that the volume occupied by 1 mole of a gas at standard temperature (273 K) and pressure (1 atm) is 22.4 dm^3 (Section 1.1). If the pressure is expressed in atm and volume in dm^3, the unit of R is atm dm^3 K^{-1} mol^{-1}, as shown below.

$$pV = nRT$$

$$R = \frac{pV}{nT} = \frac{1 \text{ atm} \times 22.4 \text{ dm}^3}{1 \text{ mol} \times 273 \text{ K}}$$

$$R = 0.082 \text{ atm dm}^3 \text{ K}^{-1} \text{ mol}^{-1}$$

With pressure in N m^{-2} and volume in m^3, R is expressed in J K^{-1} mol^{-1}.

$$R = \frac{pV}{nT} = \frac{101\,325 \text{ N m}^{-2} \times 0.0224 \text{ m}^3}{1 \text{ mol} \times 273 \text{ K}}$$

$$= 8.314 \text{ N m K}^{-1} \text{ mol}^{-1} = 8.314 \text{ J K}^{-1} \text{ mol}^{-1}$$

In calculations, when the pressure is expressed in atm and volume in dm^3, use the value $R = 0.082$ atm dm^3 K^{-1} mol^{-1} and if the pressure is expressed in N m^{-2} (or Pa) and volume in m^3, $R = 8.314$ J K^{-1} mol^{-1}. You need to be careful to use consistent units when using $pV = nRT$.

The **units** of volume, pressure and temperature can be a source of confusion. The SI unit of volume is m^3 and it is useful to remember that

$$1 \text{ m}^3 = 10^3 \text{ dm}^3 = 10^6 \text{ cm}^3$$

The volume of a gas is often expressed in litres (l) or in dm^3,

$$1 \text{ dm}^3 = 1 \text{ l}$$

The SI unit of pressure is newtons per square metre (N m^{-2}) – this is also known as Pascals (Pa).

$$1 \text{ N m}^{-2} = 1 \text{ Pa}$$

Standard pressure is 1 atmosphere (1 atm) and it is equivalent to 1.01325×10^5 N m^{-2}.

$$1 \text{ atm} = 1.01325 \times 10^5 \text{ N m}^{-2} = 101.325 \text{ kPa}$$

Temperature is expressed in degrees Kelvin (K). Standard temperature is 0°C (273 K).

The **relative molecular mass** of a gas can be calculated making use of the ideal gas law $pV = nRT$. Since the number of moles (n) is related to the mass (m) and the molar mass (M), substituting for n and rearranging gives an equation for M.

$$pV = nRT$$

$$\text{But } n = \frac{m}{M}$$

$$\text{So } pV = \frac{m}{M} RT$$

$$M = \frac{mRT}{pV}$$

2.2 Calculations based on Boyle's law, Charles' law and the general gas equation

Example 2.1 Calculating the volume of a gas at different temperatures and pressures

A gas occupies a volume of 120.0 cm^3 at 17°C and 0.95 atm pressure. Calculate its volume at 25°C and 1 atm.

Solution
Initial conditions:

$$p_1 = 0.95 \text{ atm}, V_1 = 120.0 \text{ cm}^3, T_1 = (17 + 273) = 290 \text{ K}$$

Final conditions:

$$p_2 = 1 \text{ atm}, \quad T_2 = (25 + 273) = 298 \text{ K}$$

$$\frac{p_1 V_1}{T_1} = \frac{p_2 V_2}{T_2}$$

$$V_2 = \frac{p_1 \times V_1 \times T_2}{T_1 \times p_2}$$

$$= \frac{0.95 \text{ atm} \times 120.0 \text{ cm}^3 \times 298 \text{ K}}{290 \text{ K} \times 1.0 \text{ atm}}$$

$$= 117.1 \text{ cm}^3$$

Note that we don't have to be fussy about units in the general gas equation, so long as the same units are used for the initial and final pressures and volumes. However, the temperatures *must* be expressed in degrees Kelvin.

Example 2.2 Calculating the pressure of a gas using $pV = nRT$

A 5 dm³ vessel contains 1.0 g of helium gas at 25°C. Calculate its pressure. [Molar mass of He = 4.0 g mol⁻¹, $R = 0.082$ atm dm³ K⁻¹ mol⁻¹]

Solution

$$\text{Moles of He } (n) = \frac{\text{Mass}}{\text{Molar mass}}$$

$$= \frac{1.0 \text{ g}}{4.0 \text{ g mol}^{-1}} = 0.25 \text{ mol}$$

$$T = (25 + 273) = 298 \text{ K}$$

$$V = 5.0 \text{ dm}^3.$$

$$p = \frac{nRT}{V}$$

$$= \frac{0.25 \text{ mol} \times 0.082 \text{ atm dm}^3 \text{ K}^{-1} \text{ mol}^{-1} \times 298 \text{ K}}{5.0 \text{ dm}^3}$$

$$= 1.22 \text{ atm}$$

In the above examples, units in the numerator and denominator have been 'cancelled out' to arrive at the final unit. In the remaining calculations, you can work out how the final unit is derived.

Example 2.3 Calculating the number of moles in a given volume of a gas

Calculate the number of moles in 8.0 dm³ of hydrogen gas at 200 kN m⁻² and 17°C. ($R = 8.314$ J K⁻¹ mol⁻¹)

Solution

$$p = 200\ 000\ \text{N m}^{-2}$$

$$V = 8\ \text{dm}^3 = 0.008\ \text{m}^3$$

$$T = (17 + 273) = 290\ \text{K}$$

$$R = 8.314\ \text{J K}^{-1}\ \text{mol}^{-1}$$

$$n = \frac{pV}{RT}$$

$$= \frac{200\ 000 \times 0.008}{8.314 \times 290} = 0.66\ \text{mol}$$

Example 2.4 Calculating the relative molecular mass of a gas

3.2 g of a gas exerts a pressure of 50.0 kN m^{-2} at 28°C in a 2.5 dm^3 container. Calculate the relative molecular mass of the gas. ($R = 8.314$ J K^{-1} mol^{-1})

Solution

$$m = 3.2\ \text{g}$$

$$p = 50\ 000\ \text{N m}^{-2}$$

$$V = 0.0025\ \text{m}^3$$

$$T = 28 + 273 = 301\ \text{K}$$

$$M = \frac{mRT}{pV}$$

$$= \frac{3.2 \times 8.314 \times 301}{50\ 000 \times 0.0025} = 64.1$$

2.3 Calculations based on Dalton's law of partial pressures

Dalton's law of partial pressures can be restated in mathematical terms as follows. For a mixture of two gases, X and Y, the total pressure is the sum of their partial pressures:

$$p_{\text{total}} = p_X + p_Y$$

Example 2.5 Calculating the total pressure of a mixture of gases

2.0 dm^3 of nitrogen at 150 kN m^{-2} pressure and 3.0 dm^3 of oxygen at 180 kN m^{-2} pressure are introduced into a 5.0 dm^3 container at the same temperature. What is the total pressure of the mixture of gases?

Solution

Step 1: Calculate the partial pressure of each gas in the 5.0 dm^3 container, using the relation $p_1V_1 = p_2V_2$ (the temperature is constant).

To calculate the partial pressure of nitrogen:

$$p_1 = 150 \text{ kN m}^{-2}, V_1 = 2.0 \text{ dm}^3 \text{ and } V_2 = 5.0 \text{ dm}^3$$

$$p_2 = \frac{p_1V_1}{V_2}$$

$$p_2(N_2) = \frac{150 \text{ kN m}^{-2} \times 2.0 \text{ dm}^3}{5.0 \text{ dm}^3}$$

$$= 60 \text{ kN m}^{-2}$$

To calculate the partial pressure of oxygen:

$$p_1 = 180 \text{ kN m}^{-2}, V_1 = 3.0 \text{ dm}^3 \text{ and } V_2 = 5.0 \text{ dm}^3$$

$$p_2(O_2) = \frac{180 \text{ kN m}^{-2} \times 3.0 \text{ dm}^3}{5.0 \text{ dm}^3}$$

$$= 108 \text{ kN m}^{-2}$$

Step 2: Add the partial pressures to get the total pressure (Dalton's law).

$$\text{Total pressure} = 60 + 108 = 168 \text{ kN m}^{-2}$$

Example 2.6 Calculating the partial pressure of a gas in a mixture

During an experiment to determine the relative atomic mass of lithium, 120 cm^3 of hydrogen gas was collected over water at 20°C and 9.9×10^4 Pa pressure. What is the pressure of dry hydrogen gas? (Water vapour pressure at 20°C $= 2.34 \times 10^3$ Pa)

Hint: When hydrogen gas is collected over water, it is saturated with water vapour. The noted pressure is the sum of the pressures exerted by the hydrogen and the water vapour.

Solution

Pressure exerted by hydrogen alone

= total pressure – water vapour pressure

$= 9.9 \times 10^4$ Pa $- 2.34 \times 10^3$ Pa

$= 9.67 \times 10^4$ Pa

Example 2.7 Calculating the pressure exerted by a mixture of gases

What is the total pressure exerted by a mixture of gases consisting of 5.6 g of CO and 1.2 g of H_2, in a 5.0 dm³ container at 1 atm pressure and 20°C? (M_R: CO = 28.0, H_2 = 2.0; R = 0.082 atm dm³ K⁻¹ mol⁻¹)

Solution

$$\text{Moles of CO} = \frac{5.6 \text{ g}}{28.0 \text{ g mol}^{-1}} = 0.2 \text{ mol}$$

$$\text{Moles of H}_2 = \frac{1.2 \text{ g}}{2.0 \text{ g mol}^{-1}} = 0.6 \text{ mol}$$

Total number of moles of CO and H_2 = 0.2 + 0.6 = 0.8 mol

$$p = \frac{nRT}{V}$$

Total pressure

$$= \frac{(0.8 \text{ mol}) \times (0.082 \text{ atm dm}^3 \text{ mol}^{-1} \text{K}^{-1}) \times (293 \text{ K})}{5.0 \text{ dm}^3}$$

$= 3.84$ atm

2.4 Practice questions

Useful data for calculations below: $R = 8.314$ J K⁻¹ mol⁻¹, 1 atm $= 1.0 \times 10^5$ N m⁻²

1. The volume occupied by a gas is 10.0 dm^3 at 15°C and 1 atm. Calculate its volume when

   (a) the pressure is doubled at constant temperature,

   (b) the pressure is lowered to 0.2 atm at constant temperature,

   (c) the temperature is raised to 50°C at constant pressure,

   (d) the temperature is raised to 20°C and the pressure to 6.0 atm.

2. A gas occupies a volume of 450.0 cm^3 at 20°C and 2.0×10^5 N m^{-2}. Calculate its volume at 0°C and 1 atm pressure.

3. The volume of a gas is 160.0 cm^3 at 27°C and 1.0×10^6 N m^{-2}. Convert this volume to volume at standard temperature (0°C) and pressure $(1.0 \times 10^5$ N m$^{-2})$.

4. The pressure exerted by a gas in a 2.5 dm^3 container is 100 kN m^{-2} at 20°C. What would be its pressure if the gas is heated to 60°C at constant volume?

5. What is the pressure exerted by 1.0 g of CO_2 gas in a 2.0 dm^3 vessel at 17°C? (M_R: $CO_2 = 44.0$)

6. 4.0 dm^3 of helium gas at 2.4 atm and 3.0 dm^3 of hydrogen gas at 0.9 atm are introduced into a 6.0 dm^3 vessel at the same temperature. What is the total pressure of the mixture of gases?

7. A 10 litre bottle contains 5.0 g of argon and 3.5 g of nitrogen at 25°C. Calculate the total pressure of the gases. (M_R: Ar $= 40.0$, $N_2 = 28.0$)

8. Calculate the number of moles in 2.4 dm^3 of butane at 17°C and 5.0 atm.

9. 3.0 g of a gas occupies a volume of 1.032 dm^3 at 100 kN m^{-2} and 25°C. Calculate the relative molecular mass of the gas.

10. 0.29 g of a liquid organic compound on vaporisation occupies 120.0 cm^3 at 25°C and 1 atm. Calculate the relative molecular mass of the compound.

3 Enthalpy changes

One-minute overview

This chapter deals with the quantitative aspects of energy changes during chemical reactions. In Section 3.1, different types of standard enthalpy changes are defined, Hess's law is explained and enthalpy diagrams and Born–Haber cycle diagrams are introduced.

Examples of calculations that follow include:
- enthalpies of formation and combustion
- enthalpy of neutralisation
- enthalpy of atomisation
- bond energy and the use of bond energies to calculate enthalpy of a reaction
- lattice energy and the Born–Haber cycle.

3.1 Basic definitions and relationships

An **exothermic reaction** is one that produces heat energy. An **endothermic reaction** is one that requires an input of heat.

Enthalpy, represented by the letter H, expresses the quantity of heat transferred at constant pressure during a reaction.

Enthalpy change, ΔH ('delta H'), is the difference between the enthalpies of the products and the enthalpies of the reactants.

$$\Delta H = H_2 - H_1$$

H_2 = sum of enthalpies of products

H_1 = sum of enthalpies of reactants.

ΔH is *negative* for an exothermic reaction and *positive* for an endothermic reaction.

A **standard enthalpy change, $\Delta H^{\ominus}$**, is the enthalpy change under standard conditions of temperature and pressure. In thermochemical calculations, standard pressure is 1 atmosphere (1 atm) and standard temperature is 298 K (25°C).

A **thermochemical equation** is a stoichiometric equation for a reaction which also states the associated enthalpy change for the reaction.

A **standard enthalpy of formation ($\Delta H_f^{\ominus}$)** is the *enthalpy change* when one mole of a substance is formed from its constituent elements at 298 K and 1 atm. For example, the enthalpy of formation of CO_2, is -393.5 kJ mol^{-1}, that is, 393.5 kJ of heat is produced when 1 mole of CO_2 is formed from C (graphite) and O_2 gas at 298 K and 1 atm. It is an exothermic reaction. The thermochemical equation for this reaction is

$$C(s)+O_2(g) \rightarrow CO_2(g), \Delta H^{\ominus} = -393.5 \text{ kJ mol}^{-1}$$

The standard enthalpy of formation of all elements is assigned to zero. For example, $\Delta H_f^{\ominus}(H_2,g) = 0.0$ kJ mol^{-1}, $\Delta H_f^{\ominus}$ (Na,s) $= 0.0$ kJ mol^{-1} $\Delta H_f^{\ominus}$ (C,graphite) $= 0.0$ kJ mol^{-1}.

Standard enthalpies of formation of some substances are given in Table 3.1.

Substance	$\Delta H_f^{\ominus}$ (kJ mol^{-1})	Substance	$\Delta H_f^{\ominus}$ (kJ mol^{-1})
$H_2O(g)$	−241.8	$CH_4(g)$	−74.8
$H_2O(l)$	−285.8	$C_2H_6(g)$	−84.7
$H_2S(g)$	−20.6	$C_2H_4(g)$	+52.2
$NH_3(g)$	−46.1	$C_2H_2(g)$	+228.0
$NO(g)$	+90.2	$HCl(g)$	−92.3
$NO_2(g)$	+33.2	$HBr(g)$	−36.4
C(diamond)	+1.9	$HI(g)$	+26.5
$CO(g)$	−110.5	$NaCl(s)$	−411.2
$CO_2(g)$	−393.5	$CaO(s)$	−635.1

Table 3.1 Enthalpies of formation of selected substances

The **standard enthalpy of combustion ($\Delta H_c^{\ominus}$)** is the *enthalpy change* when one mole of a substance is burned completely in air or oxygen at 298 K and 1 atm. For example, the standard enthalpy of combustion of methane is -890.3 kJ mol^{-1}.

$$CH_4(g)+2O_2(g) \rightarrow CO_2(g)+2H_2O(l), \Delta H^{\ominus} = -890.3 \text{ kJ mol}^{-1}$$

Standard enthalpies of combustion of some substances are given in Table 3.2.

All combustion reactions are exothermic reactions

Substance	Formula	$\Delta H^{\ominus}c$ (kJ mol^{-1})
Hydrogen	$H_2(g)$	−285.8
Carbon (graphite)	$C(s)$	−393.5
Methane	$CH_4(g)$	−890.3
Ethanol	$C_2H_5OH(l)$	−1367.3
Benzene	$C_6H_6(l)$	−3267.4
Ethyne	C_2H_2	−1300.8

Table 3.2 Standard enthalpies of combustion of selected substances ◄

The **enthalpy of neutralisation** is the *enthalpy change* when one mole of H$^+$ ions provided by a dilute acid is neutralised completely by a mole of OH$^-$ ions provided by one dilute base.

$$H^+(aq)+OH^-(aq) \rightarrow H_2O(l), \Delta H = -57.3 \text{ kJ mol}^{-1}$$

The **Enthalpy of atomisation, ΔH_{atm}**, is the amount of energy required for the formation of 1 mole of gaseous atoms from the element in its standard (physical) state. For example,

$$K(s) \rightarrow K(g), \Delta H_{atm} = +89.2 \text{ kJ mol}^{-1}$$

$$\tfrac{1}{2}Cl_2(g) \rightarrow Cl(g), \Delta H_{atm} = +121.7 \text{ kJ mol}^{-1}$$

The **enthalpy of reaction, $\Delta H^{\ominus}_{rxn}$**, in general, is the *enthalpy change* when the number of moles of the reactants in a stoichiometric equation react completely to form the products at 298 K and 1 atm pressure. For example, when 2 moles of H$_2$S gas react with 1 mole of SO$_2$ gas, 235.6 kJ of heat is produced under standard conditions.

$$2H_2S(g)+SO_2(g) \rightarrow 2H_2O(l)+3S(s), \Delta H_{rxn} = -235.6 \text{ kJ mol}^{-1}$$

The **bond energy BE, (or bond dissociation energy)** is the amount of energy *required* to break the bonds in 1 mole of gaseous diatomic molecules, to form gaseous atoms. For example,

This can be confusing. For example in this reaction you might ask – 235.6 kJ per mole of what – H$_2$S? SO$_2$? Well, it's neither of these. The unit mol^{-1} refers to the whole stoichiometric reaction as written. So, when 2 moles of H$_2$S react with 1 mole of SO$_2$ to make 2 moles of H$_2$O and 3 moles of S 235.6 kJ of energy is released.

Energy is required to break a bond and energy is evolved when a bond is formed.

$$HCl(g) \rightarrow H(g) + Cl(g), \Delta H = +431.0 \text{ kJ mol}^{-1}$$

For molecules which contain more than one bond, the **average bond energy** is the energy required to break 1 mole of a particular bond. Table 3.3 lists some average bond energies.

Table 3.3 Average bond energies ▶

Bond	BE (kJ mol⁻¹)	Bond	BE (kJ mol⁻¹)
H—H	435.9	C—H	413.0
H—F	568.0	C—F	485.0
H—Cl	432.0	C—Cl	346.0
H—Br	366.0	C—Br	290.0
H—I	298.3	C—I	228.0
F—F	158.0	C—C	347.0
Cl—Cl	243.4	C=C	612.0
Br—Br	192.9	C≡C	838.0
I—I	151.2	C—O	358.0
O—H	464.0	C=O	745.0
S—H	364.0	O=O	498.3
N—H	391.0	N≡N	945.4

3.2 Enthalpy of reaction in relation to $\Delta H_f^\ominus$ and bond energy

The relationship between enthalpy of reaction and enthalpy of formation is given by

Σ stands for 'sum of'

$$\text{enthalpy of reaction, } \Delta H_{rxn}^\ominus = \Sigma \Delta H_f^\ominus \text{ (products)} -\Sigma \Delta H_f^\ominus \text{ (reactants)}$$

The relationship between enthalpy of reaction and bond energy is given by

$$\Delta H_{rxn}^\ominus = \Sigma \text{ BE (reactants)} -\Sigma \text{ BE (products)}$$

3.3 Hess's law and enthalpy diagrams

Hess's law states that the energy evolved or absorbed during a reaction is the same whether the reaction takes place in one step or through a number of steps. For example, carbon dioxide can be formed by a direct reaction (1) or through two steps (2 and 3). The enthalpy of formation of CO_2 is −393.5 kJ mol⁻¹ whether it is formed in one step or two.

$$C(s) + O_2 (g) \rightarrow CO_2 (g), \Delta H_1 = -393.5 \text{ kJ mol}^{-1} \quad (1)$$

$$C(s) + \tfrac{1}{2}O_2 (g) \rightarrow CO (g), \Delta H_2 = -110.5 \text{ kJ mol}^{-1} \quad (2)$$

$$CO(g) + \tfrac{1}{2}O_2 (g) \rightarrow CO (g), \Delta H_3 = -283.0 \text{ kJ mol}^{-1} \quad (3)$$

Adding equations (2) and (3) gives

$$C(s) + O_2 (g) \rightarrow CO_2 (g), \quad \Delta H_1 = -393.5 \text{ kJ mol}^{-1}$$

This can be summarised as shown in Figure 3.1.

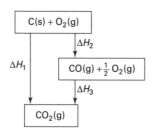

Figure 3.1
Illustration of
Hess's Law
◀

$$\Delta H_1 = \Delta H_2 + \Delta H_3$$

Hess's law enables us to add or subtract thermochemical equations like algebraic equations.

An **enthalpy diagram** is a simple energy profile of a reaction to show the enthalpies of the reactants compared with those of the products. Two examples are given below.

The enthalpy diagram for the complete combustion of carbon (the formation of carbon dioxide) is shown in Figure 3.2

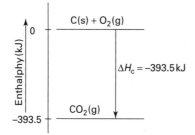

Figure 3.2
Enthalpy
diagram for the
formation of
CO_2
◀

Note that the enthalpy of a system containing only elements under standard conditions is 0 kJ and that of 1 mole of CO_2 is -393.5 kJ. Negative enthalpy change is shown by a downward arrow.

The enthalpy changes for the formation of CO_2 directly and through CO are shown in Figure 3.3

Figure 3.3 Enthalpy diagram for the direct and indirect formation of carbon dioxide ▶

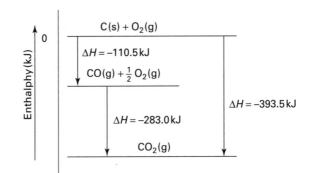

3.4 Lattice energy and the Born–Haber cycle

Lattice energy, ΔH_{lat}, is the energy evolved when 1 mole of a solid ionic compound is formed from its gaseous ions under standard conditions. For example,

$$Na^+(g) + Cl^-(g) \rightarrow NaCl(s), \Delta H = -776 \text{ kJ mol}^{-1}$$

The lattice energy of an ionic compound cannot be determined directly, but can be calculated indirectly using the **Born–Haber cycle** which consists of enthalpy levels of:

● the elements in their standard state

● the ions in the gaseous state

● the ionic solid.

It is summarised in Figure 3.4. See Examples 3.11 and 3.12.

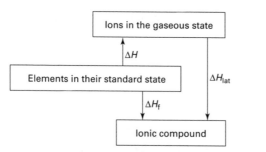

3.5 Calculations involving enthalpies of formation and combustion

Example 3.1 Calculating the heat evolved during combustion

Calculate the energy evolved when 10.0 g of ethanol, C_2H_5OH is burned completely to give CO_2 and H_2O. $(\Delta H_c^{\ominus} (C_2H_5OH) = -1367.3 \text{ kJ mol}^{-1})$

Solution

$$C_2H_5OH(l) + 3O_2(g)_2 \rightarrow 2CO_2(g) + 3H_2O(l), \ \Delta H_c^{\ominus}$$
$$= -1367.3 \text{ kJ mol}^{-1}$$

$$\text{Molar mass } (C_2H_5OH) = 46 \text{ g mol}^{-1}$$

1 mole (46 g) of ethanol on combustion produces 1367.3 kJ of heat.

$$\text{Energy evolved when 10.0 g of} \atop \text{ethanol is burned} = \frac{1367.3 \text{ kJ} \times 10.0 \text{ g}}{46 \text{ g}}$$
$$= 297.2 \text{ kJ}$$

Example 3.2 Calculating how much fuel to use in combustion

What mass of butane, C_4H_{10} should be burned to produce 1000 kJ of heat? $(\Delta H_c^{\ominus}(C_4H_{10}) = -2876.5 \text{ kJ mol}^{-1})$

Solution

$$C_4H_{10}(g) + 6\tfrac{1}{2}O_2(g) \rightarrow 4CO_2(g) + 5H_2O(l), \Delta H^{\ominus}c$$
$$= -2876.5 \text{ kJ mol}^{-1}$$

2876.5 kJ of heat is produced by the combustion of 1 mole (58 g) of butane.

mass of butane needed to produce 1000 kJ of heat

$$= \frac{58 \text{ g} \times 1000 \text{ kJ}}{2876.5 \text{ kJ}}$$

$$= 20.2 \text{ g of butane}$$

Example 3.3 Calculating enthalpy of reaction using enthalpies of formation

Using the enthalpies of formation in Table 3.1, calculate the enthalpy change of the reaction

$$2NO(g) + O_2(g) \rightarrow 2NO_2(g)$$

Solution

$$\Delta H^{\ominus}_{rxn} = \Sigma \Delta H^{\ominus}_{f} (\text{products}) - \Sigma \Delta H^{\ominus}_{f} (\text{reactants})$$

$$\Delta H^{\ominus} = [2\Delta H^{\ominus}_{f} (NO_2, g)] - [2\Delta H^{\ominus}_{f} (NO, g) + \Delta H^{\ominus}_{f} (O_2, g)]$$

$$= (2 \times +33.2) - (2 \times +90.2 + 0)$$

$$= -114.0 \text{ kJ mol}^{-1}$$

Example 3.4 Calculating enthalpy of formation from enthalpy of combustion

Calculate the enthalpy of formation of methane (CH_4) using the data given below.

$$CH_4(g) + 2O_2(g) \rightarrow CO_2(g) + 2H_2O(l), \quad \Delta H^{\ominus}_{c}$$
$$= -890.3 \text{ kJ mol}^{-1}$$

Solution

$$\Delta H^{\ominus}_{c} = \Delta H^{\ominus}_{rxn} = \Sigma \Delta H^{\ominus}_{f} (\text{products}) - \Sigma \Delta H^{\ominus}_{f} (\text{reactants})$$

$$\Delta H^{\ominus}_{c} = [\Delta H^{\ominus}_{f} (CO_2, g) + 2\Delta H^{\ominus}_{f} (H_2O, l)]$$
$$- [\Delta H^{\ominus}_{f} (CH_4, g) + 2\Delta H^{\ominus}_{f} (O_2, g)]$$

$$-890.3 = [-393.5 + (2 \times -285.8)] - [\Delta H^{\ominus}_{f} (CH_4, g) + 0]$$

$$\Delta H^{\ominus}_{f} (CH_4, g) = -74.8 \text{ kJ mol}^{-1}$$

3.6 Calculations involving enthalpy of neutralisation

Example 3.5 Calculating enthalpy of neutralisation

50 cm^3 of a 0.12 mol dm^{-3} solution of hydrochloric acid is neutralised completely by a solution of sodium hydroxide and produces 347.2 J of heat. Calculate the molar enthalpy of neutralisation.

Solution

0.12 mole of HCl is present in 1 dm^3 (1000 cm^3) of acid.

$$\text{Number of moles of HCl in } 50 \text{ cm}^3 \text{ of } 0.12 \text{ mol dm}^{-3} \text{ soln} = \frac{0.12 \text{ mol} \times 50 \text{ cm}^3}{1000 \text{ cm}^3}$$
$$= 0.006 \text{ mol}$$

Heat evolved when 0.006 mol of HCl
is neutralised by NaOH $= 347.2$ J

$$\text{Heat evolved when 1 mol of HCl is neutralised competely} = \frac{347.2 \text{ J} \times 1 \text{ mol}}{0.006 \text{ mol}}$$
$$= 57866.6 \text{ J}$$

Molar enthalpy of neutralisation $= -57.87$ kJ mol^{-1}

Example 3.6

50.0 cm^3 of 1.0 mol dm^{-3} HCl is added to an equal volume of NaOH solution of the same concentration in a polystyrene cup, both solutions at an initial temperature of 20°C. Assuming that there is no heat loss to the surroundings, calculate the rise in temperature of the mixture.

Assume that the specific heat capacity of solution = 4.2 J g^{-1} K^{-1}, density of solution = 1.0 g cm^{-3} and $\Delta H^{\ominus}_{\text{neutralisation}} = -57.3$ kJ mol^{-1}.

Solution

Step 1: Calculate the number of moles of HCl in 50 cm^3 of 1.0 mol dm^{-3} solution.

$$\text{Number of moles of HCl in } 50 \text{ cm}^3 \text{ of } 1.0 \text{ mol dm}^{-3} \text{ soln} = \frac{1.0 \text{ mol} \times 50 \text{ cm}^3}{1000 \text{ cm}^3}$$

$$= 0.05 \text{ mol}$$

Step 2: Calculate the energy evolved when 0.05 mol of HCl is neutralised by NaOH.

Energy evolved when 1 mol of HCl is neutralised by NaOH = 57.3 kJ

$$\text{Heat evolved by neutralising } 0.05 \text{ mol of HCl} = 57300 \text{ J} \times 0.05$$
$$= 2865 \text{ J}$$

Step 3: Use the equation, $Q = mc\Delta t$, (where Q is the quantity of energy, m is the mass of the solution, c is the specific heat capacity of the solution and Δt is the rise in temperature).

$$Q = 2865 \text{ J}$$
$$m = 100 \text{ g}$$
$$c = 4.2 \text{ J g}^{-1} \text{ K}^{-1}$$
$$2865 \text{ J} = 100 \text{ g} \times 4.2 \text{ J g}^{-1} \text{ K}^{-1} \times \Delta t \text{ K}$$

$$\Delta t = \frac{2865 \text{ J}}{100 \text{ g} \times 4.2 \text{ J g}^{-1} \text{ K}^{-1}}$$
$$= 6.8 \text{ K} = 6.8°\text{C}$$

3.7 Calculations using bond energies

The bond energies given in Table 3.3 are used in the following calculations, unless otherwise stated.

Example 3.7 Calculating the total bond energy in a substance

Calculate the total bond dissociation energy of C_2H_6.

Hint: One molecule of C_2H_6 contains one C – C bond and six C – H bonds

Solution

Total bond dissociation energy of C_2H_6
$$= \text{BE } (C - C) + 6 \times \text{BE}(C - H)$$
$$= 347.0 + (6 \times 413.0)$$
$$= 2825.0 \text{ kJ}$$

Example 3.8 Calculating average bond energy

Calculate the average bond energy of a C – H bond in methane.

$$CH_4(g) \rightarrow C(g) + 4H(g), \Delta H = + 1664.0 \text{ kJ}$$

Solution

$$BE (C - H) = \frac{1664.0 \text{ kJ}}{4} = 416.0 \text{ kJ mol}^{-1}$$

> **Hint:** In the reaction, four C – H bonds are broken.

Example 3.9 Calculating enthalpy of reaction using bond energies

Calculate the enthalpy change of the following reaction.

$$CH_4(g) + 2O_2(g) \rightarrow CO_2(g) + 2H_2O(g)$$

[Given: BE (C = O in CO_2) = 805.0 kJ mol^{-1},
BE (C – H in CH_4) = 416.0 kJ mol^{-1}]

Solution

Bonds in the reactants: 4 C – H bonds and 2 O = O bonds

Bonds in the products: 2 C = O bonds and 4 H – O bonds

$\Delta H_{rxn} = \Sigma$ BE (reactants) $- \Sigma$ BE (products)

$= [4 \times (C - H) + 2 \times (O = O)] - [2 \times (C = O)$
$+ 4 \times (H - O)]$

$= [(4 \times 416.0) + (2 \times 498.3)] - [(2 \times 805.0)$
$+ (4 \times 464.0)]$

$= -805.4 \text{ kJ mol}^{-1}$

> Table 3.3 gives a value of 745.0 kJ for the C = O bond. But remember that this table gives *average* bond energies, calculated over a range of different compounds. This particular C = O bond is the bond in CO_2 molecules

Example 3.10 Calculating enthalpy of reaction using bond energies

Calculate the enthalpy change of the reaction between ethene and bromine.

$$H_2C = CH_2(g) + Br_2(g) \rightarrow CH_2Br - CH_2Br(g)$$

Solution

Hint: Draw the structural formulae to see all the bonds in the reactant and product molecules.

$$\begin{array}{c} H \\ \diagdown \\ C = C \\ \diagup \\ H \end{array} \begin{array}{c} H \\ \diagup \\ \diagdown \\ H \end{array} + Br{-}Br \longrightarrow \begin{array}{c} H \ \ H \\ | \ \ | \\ H{-}C{-}C{-}H \\ | \ \ | \\ Br \ \ Br \end{array}$$

Bonds in the reactants: $4C - H$, $1C = C$ and $1Br - Br$

Bonds in the products: $4C - H$, $1C - C$ and $2C - Br$

$\Delta H_{rxn} = \Sigma$ BE (reactants) $- \Sigma$ BE (products)

$= [4 \times (C - H) + (C = C) + (Br - Br)] - [4 \times (C - H)$
$+ (C - C) + 2 \times (C - Br)]$

$= [(4 \times 413.0) + 612.0 + 192.9] - [(4 \times 413.0) + 347.0$
$+ (2 \times 290.0)]$

$= -122.1$ kJ

3.8 Calculations involving lattice energy and the Born–Haber cycle

The lattice energy of an ionic compound can be calculated using the enthalpy of atomisation (ΔH_{atm}) of the elements, the ionization energy of the metal (ΔH_{IE}), the electron affinity of the non-metal (ΔH_{EA}) and the enthalpy of formation of the compound (ΔH_f).

Example 3.11 Calculating lattice energy

Draw a Born–Haber cycle diagram for the formation of NaCl(s) using the following thermochemical data and calculate the lattice energy of sodium chloride.

$$Na(s) \rightarrow Na(g), \qquad \Delta H_{atm} = +108 \text{ kJ mol}^{-1} \ (1)$$

$$\tfrac{1}{2}Cl_2(s) \rightarrow Cl(g), \qquad \Delta H_{atm} = +122 \text{ kJ mol}^{-1} \ (2)$$

$$Na(g) \rightarrow Na^+(g) + e, \ \Delta H_{IE} = +495 \text{ kJ mol}^{-1} \ (3)$$

$$Cl(g) + e^- \rightarrow Cl^-(g), \qquad \Delta H_{EA} = -360 \text{ kJ mol}^{-1} \ (4)$$

$$Na(s) + \tfrac{1}{2}CL_2(g) \rightarrow NaCl(s), \qquad \Delta H_f^{\ominus} = -411 \text{ kJ mol}^{-1} \ (5)$$

Solution

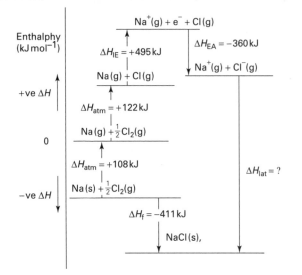

Figure 3.5
Born–Haber
cycle to
calculate ΔH_{lat}
(NaCl)
◄

From the diagram, it can be seen that the lattice energy, ΔH_{lat} can be obtained by subtracting 360 from the sum of 495, 122, 108 and 411. This will give the magnitude of the change. Since the formation of a lattice is exothermic, ΔH_{lat} takes a negative sign.

$$\Delta H_{lat} = -(495 + 122 + 108 + 411 - 360) = -776 \text{ kJ mol}^{-1}$$

Alternative method using Hess's law

Reverse equations (1), (2), (3) and (4) and add the reversed equations and equation (5). Remember that when an equation is reversed, the sign of ΔH also changes.

$$Na(g) \rightarrow Na(s), \qquad \Delta H = -108 \text{ kJ mol}^{-1}$$

$$Cl(g) \rightarrow \tfrac{1}{2}Cl_2(g), \qquad \Delta H = -122 \text{ kJ mol}^{-1}$$

$$Na^+(g) + e^- \rightarrow Na(g) \qquad \Delta H = -495 \text{ kJ mol}^{-1}$$

$$Cl^-(g) \rightarrow Cl(g) + e^- \qquad \Delta H = +360 \text{ kJ mol}^{-1}$$

$$Na(s) + \tfrac{1}{2}Cl_2(g) \rightarrow NaCl(s), \qquad \Delta H_f = -411 \text{ kJ mol}^{-1}$$

$$\overline{Na^+(g) + Cl^-(g) \rightarrow NaCl(s), \qquad \Delta H_{lat} = -766 \text{ kJ mol}^{-1}}$$

When the equations are added, identical substances on opposite sides of the equations cancel each other out to give the resultant equation. For example, Na(s) in the first equation cancels out Na(s) in the fifth equation.

Example 3.12 Calculating enthalpy of formation

Draw a Born–Haber cycle for the formation of potassium chloride and use the cycle to calculate the enthalpy of formation of KCl(s).

Enthalpy of atomisation of potassium, $\Delta H^{\ominus}_{atm} = +89$ kJ mol^{-1}
Enthalpy of atomisation of chlorine, $\Delta H^{\ominus}_{atm} = +122$ kJ mol^{-1}
Ionization energy of potassium, $\Delta H^{\ominus}_{IE} = +419$ kJ mol^{-1}
Electron affinity of chlorine, $\Delta H^{\ominus}_{EA} = -360$ kJ mol^{-1}
Lattice energy of potassium chloride, $\Delta H^{\ominus}_{lat} = -711$ kJ mol^{-1}

Solution

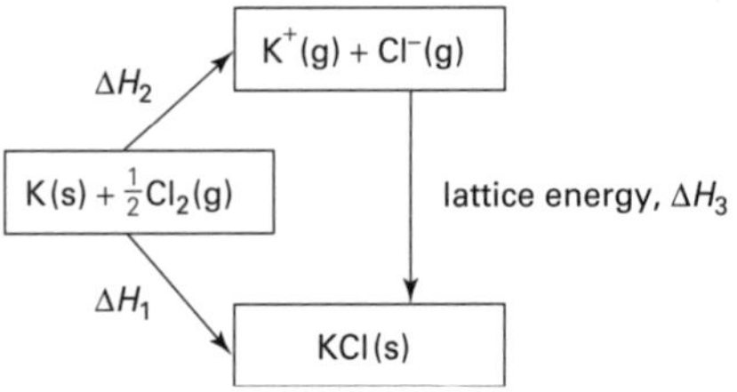

▶
*Figure 3.6
Born–Haber
cycle for the
formation of
KCl (s)*

ΔH_2 → K$^+$(g) + Cl$^-$(g)

K(s) + $\frac{1}{2}$Cl$_2$(g)   lattice energy, ΔH_3

ΔH_1

KCl(s)

ΔH_2 condenses several steps in the full Born–Haber cycle

ΔH_1 = the enthalpy of formation of KCl(s)

ΔH_2 = total enthalpy change for the formation of gaseous K$^+$ ions and Cl$^-$ ions

$$= \Delta H_{atm} (K) + \Delta H_{atm} (\tfrac{1}{2}Cl_2) + \Delta H_{IE} (K) + \Delta H_{EA} (Cl)$$

$$= 89 + 122 + 419 - 360 = +270 \text{ kJ}$$

$\Delta H_3 = -711$ kJ

$\Delta H_1 = \Delta H_2 + \Delta H_3$ $= +270 - 711$ $= -441$ kJ

3.9 Practice questions

1. State which of the following reactions are exothermic and which are endothermic.

   (a)   Br$_2$(g) → 2Br (g) $\Delta H = +193$ kJ mol^{-1}

   (b)   F(g) → F$^+$ (g) + e$^-$ $\Delta H = +1681$ kJ mol^{-1}

(c) $F(g) + e^- \rightarrow F^-(g)$ $\Delta H = -328$ kJ mol^{-1}

(d) $Fe_2O_3(s) + 3Cu(s) \rightarrow 2Fe(s) + 3CuO(s)$

$\Delta H = +352$ kJ mol^{-1}

(e) $Na^+(g) + Cl^-(g) \rightarrow NaCl(s)$, $\Delta H = -776$ kJ mol^{-1}

2. $2CH_3OH(l) + 3O_2(g) \rightarrow 2CO_2(g) + 4H_2O(l)$,

$\Delta H = -1452.8$ kJ mol^{-1}

(a) What is the heat of combustion, ΔH_c, of methanol?

(b) Calculate the amount of heat produced when 5.0 g of methanol is burned completely in air to form CO_2 and H_2O. (M_R: $CH_3OH = 32.0$)

3. Use the following enthalpies of combustion

$C(s) + O_2(g) \rightarrow CO_2(g)$, $\Delta H = -393.5$ kJ mol^{-1}

$H_2(g) + \tfrac{1}{2}O_2(g) \rightarrow H_2O(l)$, $\Delta H = -285.8$ kJ mol^{-1}

$C_2H_4(g) + 3O_2(g) \rightarrow 2CO_2(g), + 2H_2O(l)$,

$\Delta H = -1410.8$ kJ mol^{-1}

to calculate the enthalpy of formation of ethene,

$2C(s) + 2H_2(g) \rightarrow C_2H_4(g)$

4. Calculate the enthalpy of formation of benzene,

$6C(s) + 3H_2(g) \rightarrow C_6H_6(l)$

given the enthalpies of combustion

$C(s) + O_2(g) \rightarrow CO_2(g)$, $\Delta H = -393.5$ kJ mol^{-1}

$H_2(g) + \tfrac{1}{2}O_2(g) \rightarrow H_2O(l)$, $\Delta H = -285.8$ kJ mol^{-1}

$C_6H_6(l) + 7\tfrac{1}{2}O_2(g) \rightarrow 6CO_2(g) + 3H_2O(l)$,

$\Delta H = -3267.4$ kJ mol^{-1}

5. Calculate the enthalpy of combustion of propane,

$C_3H_8(g) + 5O_2(g) \rightarrow 3CO_2(g) + 4H_2O(l)$

given the enthalpies of formation

$\Delta H_f(CO_2, g) = -393.5$ kJ mol^{-1}

$$\Delta H_f \, (H_2O, l) = -285.8 \text{ kJ mol}^{-1}$$

$$\Delta H_f \, (C_3H_8, g) = -104.5 \text{ kJ mol}^{-1}$$

6. Calculate the enthalpy of decomposition of copper(II) hydroxide, $Cu(OH)_2$.

$$Cu(OH)_2(s) \rightarrow CuO(s) + H_2O(l)$$

using the enthalpies of formation

$$\Delta H_f \, (Cu(OH)_2, s) = -450.0 \text{ kJ mol}^{-1}$$

$$\Delta H_f \, (CuO, s) \quad = -157.0 \text{ kJ mol}^{-1}$$

$$\Delta H_f \, (H_2O, l) \quad = -285.8 \text{ kJ mol}^{-1}$$

7. Calculate the enthalpy change of the reaction between methane and chlorine to form chloromethane and hydrogen chloride. (Bond energies in kJ mol^{-1}: C – H = 413.0, Cl – Cl = 243.4, C – Cl = 346.0, H – Cl = 432.0)

$$CH_4(g) + Cl_2(g) \rightarrow CH_3Cl(g) + HCl(g)$$

8. Calculate the average C – C bond energy in propane, given that

$$BE(C - H) \text{ in } C_3H_8 = 413.0 \text{ kJ.}$$

$$C_3H_8(g) \rightarrow 3C(g) + 8H(g), \Delta H = +3998.0 \text{ kJ mol}^{-1}$$

9. Calculate the enthalpy change of the reaction given below. (Bond energies in kJ mol^{-1}: C – H = 413.0, C – C = 347.0, C – O = 358.0, O – H = 464.0, H – Cl = 432.0, C – Cl = 346.0)

$$
\begin{array}{ccc}
\text{H} \quad \text{H} & & \text{H} \quad \text{H} \\
| \quad\; | & & | \quad\; | \\
\text{H}-\text{C}-\text{C}-\text{O}-\text{H}\,(g) + \text{H}-\text{Cl}(g) & \longrightarrow & \text{H}-\text{C}-\text{C}-\text{Cl}(g) + \text{H}_2\text{O}\,(g) \\
| \quad\; | & & | \quad\; | \\
\text{H} \quad \text{H} & & \text{H} \quad \text{H}
\end{array}
$$

10. Using the I – I and Cl – Cl bond energies and the enthalpy change of the reaction between iodine and chlorine gases (given below) calculate the bond energy

of the I – Cl bond in ICl_3. (Bond energies in kJ mol^{-1}: I – I = 151.2, Cl – Cl = 243.4)

$$I_2(g) + 3Cl_2(g) \rightarrow 2ICl_3(g), \ \Delta H = -27.4 \text{ kJ mol}^{-1}$$

11. Using the data given below calculate the lattice energy of CaO(s).

$$Ca(s) \rightarrow Ca(g) \qquad \Delta H_{atm} = +178.2 \text{ kJ mol}^{-1}$$

$$Ca(g) \rightarrow Ca^{2+}(g) + 2e^- \quad \Delta H = +1735.0 \text{ kJ mol}^{-1}$$

$$\tfrac{1}{2}O_2(g) \rightarrow O(g) \qquad \Delta H_{atm} = +249.2 \text{ kJ mol}^{-1}$$

$$O(g) + 2e^- \rightarrow O^{2-}(g) \qquad \Delta H = +657.0 \text{ kJ mol}^{-1}$$

$$Ca(s) + \tfrac{1}{2}O_2(g) \rightarrow CaO(s) \qquad \Delta H_f = -635.1 \text{ kJ mol}^{-1}$$

12. Using the data given below calculate the standard enthalpy of formation of $MgI_2(s)$.

$$Mg(s) \rightarrow Mg(g), \qquad \Delta H_{atm} = +148 \text{ kJ mol}^{-1}$$

$$Mg(g) \rightarrow Mg^+(g) + e^-, \Delta H_{IE1} = +738 \text{ kJ mol}^{-1}$$

$$Mg^+(g) \rightarrow Mg^{2+}(g) + e^- \ \Delta H_{IE2} = +1451 \text{ kJ mol}^{-1}$$

$$\tfrac{1}{2}I_2(s) \rightarrow I(g), \qquad \Delta H_{atm} = +107 \text{ kJ mol}^{-1}$$

$$I(g) + e^- \rightarrow I^-(g), \qquad \Delta H_{EA} = -295 \text{ kJ mol}^{-1}$$

$$Mg^{2+}(g) + 2I^- \rightarrow MgI_2(s), \qquad \Delta H_{lat} = -2325 \text{ kJ mol}^{-1}$$

4   **Chemical equilibrium**

One-minute overview

This chapter deals with reversible chemical reactions and the equilibrium constants, K_c and K_p. Most chemical reactions are reversible, at least to some extent. A system containing the reactants and products of a reversible reaction is at equilibrium when the rate of the forward reaction equals that of the reverse reaction. The concentrations of reactants and products in an equilibrium system remain constant and depend on the amounts of the substances started with, the equilibrium constant and temperature.

The illustrated examples demonstrate:
- the calculation of the equilibrium constants K_c and K_p
- how to deduce units of equilibrium constants
- the calculation of concentrations of substances present in an equilibrium mixture.

4.1 Important terms and relationships

A **chemical equilibrium** is a dynamic equilibrium in which a forward reaction and its reverse reaction take place simultaneously at equal rates.

The **factors** that affect a system at equilibrium are the concentrations of the substances, the temperature and the pressure (in case of a gaseous system). Note that a catalyst has no effect on a system at equilibrium.

The **equilibrium constant** K_c for a reaction between A and B to form C and D can be expressed as follows.

$$a\text{A} + b\text{B} \rightleftharpoons c\text{C} + d\text{D}$$

$$K_c = \frac{[C]^c \, [D]^d}{[A]^a \, [B]^b}$$

The terms a, b, c and d are the number of molecules of A, B, C and D respectively in the stoichiometric equation, and [ ] stands for a concentration in mol dm^{-3}. K_c is a constant for a given reaction at a constant temperature. For example, for the reaction between nitrogen and hydrogen to form ammonia,

$$N_2(g) + 3H_2(g) \rightleftharpoons 2NH_3(g)$$

has the equilibrium expression

$$K_c = \frac{[NH_3]^2}{[N_2] \, [H_2]^3}$$

The **equilibrium constant** K_p is used for gaseous systems when the amounts of the substances are expressed in partial pressures. For example,

$$N_2(g) + 3H_2(g) \rightleftharpoons 2NH_3(g)$$

$$K_p = \frac{p_{NH_3}^2}{p_{N_2} \times p_{H_2}^3}$$

The partial pressure, p, of a gaseous substance is in atm or in N m^{-2} and K_p is the constant for a particular reaction at constant temperature. The **partial pressure** of a gas (Section 2.1) can be calculated by multiplying its mole fraction by the total pressure of the gaseous mixture.

The value of the equilibrium constant for a reaction is altered by a **change in temperature**. *As the temperature increases, the value of the equilibrium constant decreases for an exothermic reaction and increases for an endothermic reaction.*

4.2 Writing expressions for equilibrium constants

Example 4.1 Writing expressions for K_c

Write an expression for the equilibrium constant K_c, and derive its unit, for each of the following reactions.

(a) $\qquad 2SO_2(g) + O_2(g) \rightleftharpoons 2SO_3(g)$

(b) $CH_3COOH(l) + C_2H_5OH(l) \rightleftharpoons CH_3COOC_2H_5(l)$
$\qquad\qquad\qquad + H_2O(l)$

Solution

(a) $K_c = \dfrac{[SO_3]^2}{[SO_2]^2\,[O_2]} \quad \dfrac{(mol\,dm^{-3})^2}{(mol\,dm^{-3})^2 \times mol\,dm^{-3}} = mol^{-1}\,dm^3$

(b) $K_c = \dfrac{[CH_3COOC_2H_5][H_2O]}{[CH_3COOH]\,[C_2H_5OH]} \quad \dfrac{mol\,dm^{-3} \times mol\,dm^{-3}}{mol\,dm^{-3} \times mol\,dm^{-3}}$

$\qquad$ = no units

Example 4.2 Writing expressions for K_p

Write an expression for the equilibrium constant K_p, and derive its unit, for each of the following reactions.

(a) $\qquad CO_2(g) + H_2(g) \rightleftharpoons CO(g) + H_2O(g)$

(b) $\qquad PCl_3(g) + Cl_2(g) \rightleftharpoons PCl_5(g)$

Solution

(a) $\qquad K_p = \dfrac{p_{CO} \times p_{H_2O}}{p_{CO_2} \times p_{H_2}} \quad \dfrac{atm \times atm}{atm \times atm} = $ no units

(b) $\qquad K_p = \dfrac{p_{PCl_5}}{p_{PCl_3} \times p_{Cl_2}} \quad \dfrac{atm}{atm \times atm} = atm^{-1}$

Example 4.3

For the reaction

$$N_2O_4(g) \rightleftharpoons 2NO_2(g)$$

(a) write an expression for the equilibrium constant K_p, and

(b) work out the unit for K_p if the partial pressures are given in N m^{-2}.

Solution

(a)
$$K_p = \frac{p_{NO_2}^{2}}{p_{N_2O_4}}$$

(b)
$$\frac{(N\ m^{-2})^2}{N\ m^{-2}} = N\ m^{-2}$$

4.3 Calculation of equilibrium constant values

Example 4.4 Calculating equilibrium constants for forward and reverse reactions

At 520 K, a 5.0 dm^3 reaction vessel contains 2.0 mole of PCl$_5$, 0.8 mole of PCl$_3$ and 0.5 mole of Cl$_2$ at equilibrium. Calculate the equilibrium constant K_{c1} for the dissociation of PCl$_5$, and the equilibrium constant K_{c2} for the formation of PCl$_5$.

Solution

Step 1: Write the equation for the dissociation of PCl$_5$ and the expression for the equilibrium constant K_{c1} for the dissociation.

$$PCl_5(g) \rightleftharpoons PCl_3(g) + Cl_2(g)$$

$$K_{c1} = \frac{[PCl_3][Cl_2]}{[PCl_5]}$$

Step 2: Calculate the equilibrium concentrations of PCl$_5$, PCl$_3$ and Cl$_2$ in mol dm^{-3}.

$$[PCl_5] = \frac{2.0\ mol}{5.0\ dm^3} = 0.4\ mol\ dm^{-3}$$

$$[PCl_3] = \frac{0.8\ mol}{5.0\ dm^3} = 0.16\ mol\ dm^{-3}$$

$$[Cl_2] = \frac{0.5 \text{ mol}}{5.0 \text{ dm}^3} = 0.1 \text{ mol dm}^{-3}$$

Step 3: Substitute the equilibrium concentrations of the substances into the expression for K_{c1} and calculate its value.

$$K_{c1} = \frac{[PCl_3][Cl_2]}{[PCl_5]} = \frac{0.16 \text{ mol dm}^{-3} \times 0.1 \text{ mol dm}^{-3}}{0.4 \text{ mol dm}^{-3}}$$

$$= 0.04 \text{ mol dm}^{-3}$$

Step 4: Write an equation for the formation of PCl_5 and the expression for the equilibrium constant K_{c2}. Calculate the value of K_{c2}.

$$PCl_3(g) + Cl_2(g) \rightleftharpoons PCl_5(g)$$

$$K_{c2} = \frac{[PCl_5]}{[PCl_3][Cl_2]} = \frac{0.4 \text{ mol dm}^{-3}}{0.16 \text{ mol dm}^{-3} \times 0.1 \text{ mol dm}^{-3}}$$

$$= 25.0 \text{ mol}^{-1} \text{ dm}^3$$

The equilibrium constant for the forward reaction, K_{c1} is the reciprocal of the equilibrium constant for the reverse reaction, K_{c2}.

Example 4.5 Calculating the value of equilibrium constants K_c

1.0 mole of PCl_5 was heated in a 1.0 dm³ flask at 520 K until equilibrium was reached. 0.82 mole of PCl_5 remained unreacted. Calculate the equilibrium concentrations of PCl_3 and Cl_2 and the equilibrium constant K_c for the dissociation of PCl_5.

Solution

$$PCl_5(g) \rightleftharpoons PCl_3(g) + Cl_2(g)$$

Moles of PCl_5 dissociated = (1.0 − 0.82) = 0.18 mol

	$PCl_5(g)$	$PCl_3(g)$	$Cl_2(g)$	
Initial amount	1.0	0	0	mol dm⁻³
Unreacted	0.82	–	–	mol dm⁻³
Formed	–	0.18	0.18	mol dm⁻³
Equilibrium concentration	0.82	0.18	0.18	mol dm⁻³

0.18 mol of PCl_5 dissociates to give 0.18 mol of PCl_3 and 0.18 mol of Cl_2.

$$K_c = \frac{[PCl_3]\,[Cl_2]}{[PCl_5]} = \frac{0.18\ mol\ dm^{-3} \times 0.18\ mol\ dm^{-3}}{0.82\ mol\ dm^{-3}}$$

$$= 0.040\ mol\ dm^{-3}$$

Example 4.6

Ethyl ethanoate ($CH_3COOC_2H_5$) is hydrolysed in the presence of hydrochloric acid to form ethanoic acid (CH_3COOH) and ethanol (C_2H_5OH). A mixture containing 5.0 g of ethyl ethanoate and 10.0 cm^3 of 1 mol dm^{-3} HCl was allowed to reach equilibrium at 20°C. The reaction mixture was found to contain 0.043 mol ethanoic acid. Calculate the equilibrium constant for the hydrolysis of ethyl ethanoate.

(M_R: H_2O = 18.0, $CH_3COOC_2H_5$ = 88.0. Assume that 10 cm^3 of hydrochloric acid contains 10 g of water.)

$$CH_3COOC_2H_5(l) + H_2O(l) \overset{H^+}{\rightleftharpoons} CH_3COOH(l) + C_2H_5OH(l)$$

Solution

Step 1: Calculate the initial amounts of ethyl ethanoate and water.

$$CH_3\ COOC_2H_5\ used = \frac{5.0\ g}{88.0\ g\ mol^{-1}} = 0.057\ mol$$

$$H_2O\ used = \frac{10.0\ g}{18.0\ g\ mol^{-1}} = 0.556\ mol$$

Step 2: Write an expression for the equilibrium constant.

$$K_c = \frac{[CH_3COOH][C_2H_5OH]}{[CH_3COOC_2H_5]\,[H_2O]}$$

Step 3: During hydrolysis, equal number of moles of ethanoic acid and ethanol are formed. The reaction mixture contains 0.043 mol of ethanoic acid and therefore 0.043 mol of ethanol. That means that 0.043 mol of each of the reactants reacts to give 0.043 mol of each of the products.

This allows us to calculate the equilibrium concentrations of all the reactants and products by dividing each by the volume of the mixture. In this case, there are two terms on the top of the expression and two on the bottom, so the volume terms world cancel out. This means that we can simply substitute the molar amounts to evaluate K_c.

	$CH_3COOC_2H_5(l)$	$H_2O(l)$	$CH_3COOH(l)$	$C_2H_5OH(l)$	
Initial amount	0.057	0.556	0	0	mol
Formed	–	–	0.043	0.043	mol
Reacted	0.043	0.043	–	–	mol
Equilibrium amounts	0.014	0.513	0.043	0.043	mol

$$K_c = \frac{0.043 \text{ mol} \times 0.043 \text{ mol}}{0.014 \text{ mol} \times 0.513 \text{ mol}} = 0.257$$

Example 4.7 Calculating values for equilibrium constants K_p

0.15 mole of $N_2O_4(g)$ was introduced into a $1\,dm^3$ vessel at 300 K. The total pressure at equilibrium was found to be 424 kNm^{-2}. Calculate the value of K_p for the dissociation of $N_2O_4(g)$ to $NO_2(g)$ at this temperature. (R = 8.31 J K^{-1})

$$N_2O_4(g) \rightleftharpoons 2NO_2(g)$$

Solution

Step 1: Work out the number of moles of the substances at equilibrium.

	$N_2O_4(g)$	$NO_2(g)$	
Initial amount	0.15	0	mol dm^{-3}
Reacted	x	–	mol dm^{-3}
Formed	–	$2x$	mol dm^{-3}
Equilibrium concentration	$(0.15 - x)$	$2x$	mol dm^{-3}

Step 2: Calculate the total number of moles of N_2O_4 and NO_2 at equilibrium.

$$\text{Total number of moles at equilibrium} = (0.15 - x) + 2x$$
$$= (0.15 + x) \text{ mol}$$

Step 3: The total pressure of the mixture of gases at equilibrium (given) is proportional to the total number of moles. Substitute the values in the equation $pV = nRT$ and solve for x.

$$\text{Total pressure} = 424 \text{ kN m}^{-2}$$
$$= 424\,000 \text{ N m}^{-2}$$
$$\text{Volume} \quad = 0.001 \text{ m}^3$$

$$pV = nRT$$

$$424\,000 \text{ N m}^{-2} \times 0.001 \text{ m}^3 = (0.15 + x) \text{ mol} \times 8.31 \text{ J K}^{-1} \text{ mol}^{-1}$$
$$\times 300 \text{ K}$$

$$x = 0.02 \text{ mol}$$

Step 4: Calculate the equilibrium partial pressures of NO_2 and N_2O_4.

$$\text{Partial pressure} = \text{total pressure} \times \text{mole fraction}$$
$$(\text{see Section 2.1})$$

$$p_{NO_2} = 424 \times \frac{2x}{(0.15 + x)}$$

$$= 99.8 \text{ kN m}^{-2}$$

$$p_{N_2O_4} = 424 \times \frac{(0.15 - x)}{(0.15 + x)}$$

$$= 324.2 \text{ kN m}^{-2}$$

Step 5: Write an expression for K_p, substitute the partial pressures and calculate the value of K_p.

$$K_p = \frac{p_{NO_2}^{\,2}}{p_{N_2O_4}}$$

$$= \frac{(99.8 \text{ kN m}^{-2})^2}{324.2 \text{ kN m}^{-2}}$$

$$= 30.7 \text{ kN m}^{-2}$$

4.4 Calculation of equilibrium concentrations

Example 4.8 Calculating the concentration of a substance in a reaction mixture

The equilibrium constant, K_c, for the reaction below is 1.85 mol^{-2} dm^6 at 623 K.

$$N_2(g) + 3H_2(g) \rightleftharpoons 2NH_3(g)$$

If a reaction mixture contains 1.2 mol dm^{-3} N_2 and 0.5 mol dm^{-3} H_2 at equilibrium, calculate the concentration of NH_3.

Solution

$$N_2(g) + 3H_2(g) \rightleftharpoons 2NH_3(g)$$

Equilibrium
concentration   1.2   0.5   ?   /mol dm^{-3}

$$K_c = \frac{[NH_3]^2}{[N_2][H_2]^3} = 1.85 \text{ mol}^{-2} \text{ dm}^6$$

$$[NH_3]^2 = 1.85 \times [N_2] \times [H_2]^3$$

$$= 1.85 \times 1.2 \times 0.5^3$$

$$= 0.2775$$

$$[NH_3] = 0.53 \text{ mol dm}^{-3}$$

Example 4.9 Calculating equilibrium concentration

A mixture of 0.1 mole of $H_2S(g)$ and 0.1 mole of $H_2(g)$ was placed in a 1 dm^3 container with excess solid sulphur. The mixture was heated and maintained at a constant temperature until equilibrium was reached. The equilibrium constant K_c for the formation of H_2S (g) is 7.0×10^{-2}. Calculate the equilibrium concentration of H_2.

Since sulphur is in the solid state, its concentration term will not appear in the expression for the equilibrium constant.

Solution

Step 1: Write an equation for the formation of H_2S and the expression for the equilibrium constant, K_c.

$$H_2(g) + S(g) \rightleftharpoons H_2S(g)$$

$$K_c = \frac{[H_2S]}{[H_2]} = 7.0 \times 10^{-2}$$

Step 2: Since the value of K_c is less than 1, $[H_2S]$ should be less than $[H_2]$. So a backward reaction takes place to produce more H_2. If x mol of H_2S dissociates to form x mol of H_2, $[H_2S] = (0.1 - x)$ mol dm^{-3} and $[H_2] = (0.1 + x)$ mol dm^{-3} at equilibrium.

$$\frac{[H_2S]}{[H_2]} = 7.0 \times 10^{-2}$$

$$\frac{(0.1 - x)}{(0.1 + x)} = 7.0 \times 10^{-2}$$

$$x = 0.087$$

$$[H_2] = 0.1 + 0.087$$
$$= 0.187 \text{ mol dm}^{-3}$$

4.5 Practice questions

1. Write an expression for the equilibrium constant K_c for each of the following reactions. Derive the unit for K_c in each case.

    (a) $\qquad Cl_2(g) + F_2(g) \rightleftharpoons 2ClF(g)$

    (b) $\qquad 2NO(g) + O_2(g) \rightleftharpoons 2NO_2(g)$

    (c) $\qquad C(s) + H_2O(g) \rightleftharpoons CO(g) + H_2(g)$

    (d) $HCOOCH_3(l) + H_2O(l) \rightleftharpoons HCOOH(l) + CH_3OH(l)$

2. Write an expression for the equilibrium constant K_p for each of the following reactions. Derive the unit for K_p in each case, assuming that the pressures were measured in atmospheres.

(a) $$N_2(g) + O_2(g) \rightleftharpoons 2NO(g)$$

(b) $$2NO_2(g) \rightleftharpoons N_2O_4(g)$$

(c) $$H_2(g) + I_2(g) \rightleftharpoons 2HI(g)$$

(d) $$2SO_3(g) \rightleftharpoons 2SO_2(g) + O_2(g)$$

3. At 1000 K, the equilibrium constant K_{p1} for the reaction below is 0.96.

$$H_2(g) + CO_2(g) \rightleftharpoons H_2O(g) + CO(g)$$

Calculate the value of the equilibrium constant K_{p2} for the reverse reaction.

$$H_2O(g) + CO(g) \rightleftharpoons H_2(g) + CO_2(g)$$

4. The partial pressures of N_2, O_2 and NO gases in an equilibrium reaction mixture are 1 atm, 1 atm and 1.2 atm respectively at 700 K. Calculate the equilibrium constant K_p for the reaction.

$$N_2(g) + O_2(g) \rightleftharpoons 2NO(g)$$

5. A 1.0 dm^3 flask contains hydrogen, iodine and hydrogen iodide at equilibrium at 500 K. If $[H_2] = [I_2] = 0.016$ mol and $[HI] = 0.2$ mol calculate the equilibrium constant for the reaction.

$$H_2(g) + I_2(g) \rightleftharpoons 2HI(g)$$

6. $$CH_3COOC_2H_5(l) + H_2O(l) \rightleftharpoons CH_3COOH(l) \\ + C_2H_5OH(l)$$

The value of equilibrium constant, K_c for the hydrolysis of ethyl ethanoate is 4.0 at 300 K. Calculate the value of the equilibrium constant for the reverse reaction.

7. A reaction vessel contains 0.97 mol dm^{-3} of SO_2, 0.91 mol dm^{-3} of O_2 and 0.19 mol dm^{-3} of SO_3 at equilibrium. Calculate the value of the equilibrium constant, K_c for the reaction

$$2SO_3(g) \rightleftharpoons 2SO_2(g) + O_2(g)$$

8. Methanal, HCHO, at a concentration of 1 mol dm^{-3} was placed in a vessel and heated to 770 K until the following equilibrium was established.

$$HCHO(g) \rightleftharpoons H_2(g) + CO(g)$$

The vessel was found to contain 0.2 mol dm^{-3} of H_2. Calculate the equilibrium constant.

9. By mixing 1 mole of ethanol and 1 mole of ethanoic acid at a particular temperature 0.67 mole of ethyl ethanoate was formed at equilibrium.

   (a) Write an equation for the reaction between ethanol and ethanoic acid.

   (b) Write an expression for the equilibrium constant.

   (c) Calculate the value of the equilibrium constant.

10. A reaction mixture was prepared with 3.0 g of ethyl ethanoate and 5.0 cm^3 of 2.0 mol dm^{-3} HCl and allowed to reach equilibrium at a particular temperature. The equilibrium mixture was found to contain 0.024 mol of ethanoic acid. Calculate the value of the equilibrium constant, K_c for the hydrolysis of ethyl ethanoate. Assume that 5.0 cm^3 of hydrochloric acid contains 5.0 g of water.

11. 0.040 mole of PCl$_5$ was shaken in a 1.0 dm^3 container and maintained at a temperature of 500 K. The total pressure of the equilibrium mixture was found to be 2.5 atm. Working to three decimal places, calculate the value of the equilibrium constant K_p for the dissociation of PCl$_5$. (R = 0.0821 dm^3 atm K^{-1} mol^{-1})

$$PCl_5(g) \rightleftharpoons PCl_3(g) + Cl_2(g)$$

Rates of reactions

One-minute overview

The study of the rates of chemical reactions is called **chemical kinetics**. In this chapter we examine some of the concepts – reaction rate, rate equation, order of reaction, reaction mechanism, molecularity of a reaction, half-life and the Arrhenius equation.

The sample calculations include:
- calculation of reaction rates
- deduction of rate equations from experimental data
- calculation of rate constants
- drawing a time–concentration graph to find half-life
- calculation of a rate constant from the half-life of a first order reaction.

5.1 Important terms

The **rate of a reaction** is the change in concentration of a substance reacted (mol dm^{-3}), or of a product formed, in unit time (usually a second). The unit for rate of reaction is mol dm^{-3} s^{-1}.

Collision theory states that reactant molecules must collide before a reaction occurs and that the rate of a reaction is proportional to the number of effective collisions (i.e. collisions which lead to the formation of an intermediate) between the reactant molecules.

The **activation energy**, E_a, is the minimum amount of energy that the reactant molecules must have in order to form an activated complex (intermediate) which then changes to give the product(s).

There are four **factors which affect reaction rates**.

- **Temperature:** at higher temperatures there are more molecules having enough energy higher than the activation energy *and* there are more frequent collisions between molecules. Both of these lead to higher reaction rates.

- **Concentration of reactants:** a higher concentration of reactants leads to more collisions between molecules which increases the reaction rate.

- **Surface area:** for reactions involving solids, if the surface area is high, more molecules come into contact with each other and so the reaction rate increases.

- **Catalyst:** a catalyst increases the rate of a reaction. Catalysts provide a different reaction pathway which requires a lower activation energy than the route without the catalyst. So, at a given temperature, there will be more molecules with enough energy to react and form the products. There are two types of catalysts – homogeneous and heterogeneous. A **homogeneous catalyst** is one which is in the same phase (physical state) as the reactants, and a **heterogeneous catalyst** is one which is in a different phase.

The **rate law** for a particular reaction describes the relationship between the rate of the reaction and the concentrations of the reactants. Consider a reaction,

$$A + B \rightarrow C$$

If the rate of reaction is proportional to [A] and also proportional to [B], then

$$\text{rate } \alpha \ [A] \ [B]$$

or

$$\textbf{rate} = k[A] \ [B]$$

The second equation is called the **rate equation**, or **rate law**, where k is the **rate constant**. This is constant for individual reactions at constant temperature. It is not necessarily so that the rate of a reaction is proportional to the concentration of each reactant. Sometimes, rate is independent of the concentration of a reactant; in other reactions, rate can be increased by a factor of 4, 9 etc. as the concentration of a reactant is increased by a factor of 2, 3 etc.

The **order of a reaction** is the sum of the exponents of the concentration terms in the rate equation. If the rate equation is, rate = k $[A]^1$ $[B]^1$, then the order of reaction = 1 + 1 = 2. The order of a reaction is commonly 0, 1 or 2 with respect to each reactant.

The order of a reaction can be deduced from a **rate–concentration** graph. For a *zero order* reaction, the rate is independent of the concentration of the reactant.

$$\text{Rate} = k \ [X]^0$$

Since $[X]^0 = 1$, then the rate is k.

For a *zero order* reaction, the rate–concentration graph is a straight line parallel to the x-axis.

For a *first order* reaction, the rate is proportional to $[X]^1$ and is a straight line graph as shown in Figure 5.1.

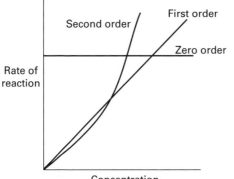

Figure 5.1 Reaction rate against concentration for zero, first and second order reactions ◄

For a *second order* reaction, the rate is proportional to $[X]^2$ and the rate–concentration graph is a curve (Figure 5.1).

The **half-life**, $t_{\frac{1}{2}}$, is the time taken for the concentration of a reactant to fall to half its original value. The half-life of a reaction can be obtained from its concentration–time graph. For a *first order reaction* the half-life period is constant. Half-life is characteristic for a particular reaction and is independent of the initial concentration of the substance. For a first order reaction, the half-life is related to the rate constant, k, by the equation

$$t_{\frac{1}{2}} = \frac{\ln 2}{k} = \frac{0.693}{k}$$

where ln is the natural logarithm.

The **Arrhenius equation** gives the relationship between the rate constant, k, and activation energy, E_a, for a reaction.

$$k = Ae^{\frac{-E_a}{RT}}$$

$$\text{or, } \ln k = \ln A - \frac{E_a}{RT}$$

$$\text{or, } \ln k = \ln A - \frac{E_a}{R} \times \frac{1}{T}$$

In the above equation, A is the Arrhenius constant, R the gas constant and T the temperature in Kelvin. A graph of $\ln k$ against $\frac{1}{T}$ is a straight line with a slope of $(-\frac{E_a}{R})$ and the intercept on the $\ln k$ axis will give the value of $\ln A$, from which A can be calculated (Figure 5.2).

Figure 5.2 Variation of rate constant with temperature ▶

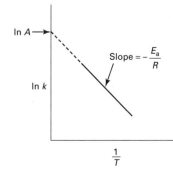

Reaction mechanism and molecularity

A reaction often proceeds by a series of steps. The **reaction mechanism** is the sequence of proposed elementary steps which lead to the formation of products. **Molecularity** is the number of chemical species reacting in an elementary step. For example, a two-step mechanism can be proposed for the dissociation of ozone, O_3, to oxygen in the presence of sunlight. The overall equation for the dissociation of ozone to oxygen is

$$2O_3 \rightarrow 3O_2$$

The proposed mechanism is

$$O_3 \rightarrow O + O_2 \quad \text{Step 1}$$

$$O + O_3 \rightarrow 2O_2 \quad \text{Step 2}$$

In Step 1, one molecule of reactant is involved in the reaction and it is called a *unimolecular reaction* (molecularity is 1). Step 2 is a *bimolecular reaction* (molecularity is 2) as two chemical species react.

Rate determining step

If a reaction takes place through a number of steps, the slowest step is the **rate determining step**. The rate equation for the reaction can be written based on the slowest step. For example, in the reaction

$$A_2 + B \rightarrow C + D$$

if the mechanism of the reaction is

$$A_2 \rightarrow X \quad \text{Step 1 slow}$$

$$X + B \rightarrow C + D \quad \text{Step 2 fast}$$

the rate determining step is the slow step (step 1) and the rate equation is,

$$\text{rate} = k[A_2].$$

It is a first order reaction. The rate of the reaction is proportional to $[A_2]$ and is independent of $[B]$.

The rate equation for a reaction cannot be deduced from the overall equation, it can only be worked out from experimental data or from the slowest step in the reaction mechanism.

5.2 Calculations involving rate of reaction, rate equations and order of reaction

Example 5.1 Calculating the rate of a reaction

For the reaction,

$$X + 2Y \rightarrow Z$$

2×10^{-4} mol dm^{-3} of Z is formed in 50 seconds. Calculate the rate of formation of Z, the rate of reaction of X and that of Y.

Hint: Since for every mole of Z formed, one mole of X and two moles of Y react, the rate of reaction of X is same as the rate of formation of Z, and the rate of reaction of Y is twice that of X.

Solution

$$\text{Rate of formation of } Z = \frac{2.0 \times 10^{-4} \text{ mol dm}^{-3}}{50 \text{ s}}$$

$$= 4.0 \times 10^{-6} \text{ mol dm}^{-3} \text{ s}^{-1}$$

$$\text{Rate of reaction of } X = 4.0 \times 10^{-6} \text{ mol dm}^{-3} \text{ s}^{-1}$$

$$\text{Rate of reaction of } Y = 2 \times 4.0 \times 10^{-6} \text{ mol dm}^{-3} \text{ s}^{-1}$$

$$= 8.0 \times 10^{-6} \text{ mol dm}^{-3} \text{ s}^{-1}$$

Example 5.2 Deducing the order of a reaction and calculating the rate constant

$$2A + B \rightarrow C$$

(a) Use the data in the table to deduce the order of reaction with respect to A, with respect to B and the overall order of the reaction.

Experi- ment	Initial [A] (mol dm^{-3})	Initial [B] (mol dm^{-3})	Initial rate of formation of C (mol dm^{-3} s^{-1})
1	0.1	0.1	1.2×10^{-4}
2	0.1	0.2	2.4×10^{-4}
3	0.2	0.2	4.8×10^{-4}

(b) Write the rate equation

(c) Calculate the value of the rate constant, k. Deduce the unit for k.

Solution

(a) Comparing Experiments 1 and 2, we can see that keeping the concentration of A constant and doubling the concentration of B doubles the rate. Therefore,

$$\text{rate} \propto [B]$$

Order with respect to B = 1

Comparing Experiments 2 and 3, keeping the concentration of B constant and doubling concentration of A, also doubles the rate. Therefore,

$$\text{rate} \propto [A]$$

Order with respect to A = 1

So the overall order of the reaction = 1 + 1 = 2

(b) The rate equation is,

$$\text{rate} = k\,[A]\,[B]$$

(c)
$$k = \frac{\text{rate}}{[A][B]}$$

Substituting the values obtained from Experiment 1,

$$k = \frac{1.2 \times 10^{-4}\ \cancel{\text{mol dm}^{-3}}\ \text{s}^{-1}}{0.1\ \cancel{\text{mol dm}^{-3}} \times 0.1\ \text{mol dm}^{-3}}$$

> You could use any of the rows in the table.

$$= 1.2 \times 10^{-2}\ \text{mol}^{-1}\ \text{dm}^3\ \text{s}^{-1}$$

Example 5.3 Calculating the initial rate of a reaction

$$H_2(g) + I_2(g) \rightarrow 2HI(g)$$

The rate equation for the reaction between hydrogen and iodine is,

$$\text{rate} = k\,[\text{H}_2]\,[\text{I}_2].$$

The rate constant, k for the reaction at 600 K is 4.4×10^{-4} mol^{-1} dm^3 s^{-1}. Calculate the initial rate of reaction between hydrogen and iodine when 0.4 mole of H_2 and 0.3 mole of I_2 are introduced into a 2 dm^3 vessel at 600 K.

Solution

$$\text{Initial }[\text{H}_2] = \frac{0.4 \text{ mol}}{2 \text{ dm}^3} = 0.2 \text{ mol dm}^{-3}$$

$$\text{Initial }[\text{I}_2] = \frac{0.3 \text{ mol}}{2 \text{ dm}^3} = 0.15 \text{ mol dm}^{-3}$$

$$\text{Rate} = k\,[\text{H}_2]\,[\text{I}_2]$$

$$= 4.4 \times 10^{-4} \text{ mol}^{-1} \text{ dm}^3 \text{ s}^{-1} \times 0.2 \text{ mol dm}^{-3}$$
$$\times 0.15 \text{ mol dm}^{-3}$$

$$= 1.32 \times 10^{-5} \text{ mol dm}^{-3} \text{ s}^{-1}$$

Example 5.4

For the reaction,

$$2\text{A}(g) + \text{B}(g) \rightarrow \text{C}(g) + \text{D}(g)$$

the rate equation is

$$\text{rate} = k\,[\text{A}]^2$$

and $k = 5.0 \times 10^{-4}$ mol^{-1} dm^3 s^{-1} at a particular temperature.

(a) Calculate the initial rate of formation of C when 0.1 mole each of A and B are mixed in a 1 dm^3 container at the above temperature.

(b) What is the initial rate formation of C when the concentrations of both A and B are doubled at the same temperature?

Solution

(a) Rate = k [A]^2

        = 5.0×10^{-4} mol^{-1} dm^3 s^{-1} × $(0.1$ mol dm$^{-3})^2$

        = 5.0×10^{-6} mol dm^{-3} s^{-1}

(b) Since the rate of the reaction is independent of the concentration of B, a change in the concentration of B does not affect the rate. As the rate is proportional to the square of the concentration of A, when the concentration of A is doubled, the rate is increased four times.

$$\text{Rate} = 4 \times 5.0 \times 10^{-6} \text{ mol dm}^{-3} \text{ s}^{-1}$$

$$= 2.0 \times 10^{-5} \text{ mol dm}^{-3} \text{ s}^{-1}$$

Example 5.5 Deducing the rate equation and calculating the rate constant

Consider the reaction,

$$H_2O_2(aq) + 2H^+(aq) + 2I^-(aq) \rightarrow 2H_2O(l) + I_2(aq)$$

(a) Can the rate equation be written from the stoichiometric equation?

(b) The following data were collected to study the rate of the above reaction.

Experi- ment	Initial $[H_2O_2]$ (mol dm^{-3})	Initial $[I^-]$ (mol dm^{-3})	Initial $[H^+]$ (mol dm^{-3})	Rate of formation of I_2 (mol dm^{-3} s^{-1})
1	0.05	0.05	0.2	1.1×10^{-5}
2	0.10	0.05	0.2	2.2×10^{-5}
3	0.05	0.05	0.3	1.1×10^{-5}
4	0.05	0.10	0.3	2.2×10^{-5}

(i) Deduce the rate equation for the reaction.

(ii) Calculate the value of the rate constant.

Solution

(a) The rate equation for a reaction cannot be deduced from the stoichiometric equation. It can only be deduced from the experimental data.

(b) (i) In experiments 1 and 2, keeping $[I^-]$ and $[H^+]$ constant and doubling $[H_2O_2]$, doubles the reaction rate. So,

$$\text{rate} \, \alpha \, [H_2O_2] \qquad (1)$$

In experiments 1 and 3, keeping $[H_2O_2]$ and $[I^-]$ constant, increasing $[H^+]$ has no effect on the reaction rate. So, the rate of reaction is independent of $[H^+]$.

In experiments 3 and 4, keeping $[H_2O_2]$ (and $[H^+]$) constant, doubling $[I^-]$ doubles the reaction rate. So,

$$\text{rate} \, \alpha \, [I^-] \qquad (2)$$

Combining (1) and (2), we can write

$$\text{rate} \, \alpha \, [H_2O_2] \, [I^-]$$

The rate equation is, therefore

$$\text{rate} = k \, [H_2O_2] \, [I^-]$$

(ii)

$$k = \frac{\text{rate}}{[H_2O_2][I^-]}$$

Substituting the values from Experiment 1

$$k = \frac{1.1 \times 10^{-5} \, \text{mol dm}^{-3} \, \text{s}^{-1}}{0.05 \, \text{mol dm}^{-3} \times 0.05 \, \text{mol dm}^{-3}}$$

$$= 4.4 \times 10^{-3} \, \text{mol}^{-1} \, \text{dm}^3 \, \text{s}^{-1}$$

5.3 Calculations involving half-life and rate constants

Example 5.6 Calculating the rate constant from the half-life

The decomposition of nitrogen(I) oxide (N_2O) is a first order reaction. The half-life of the reaction is 0.9 s at 1000 K. Calculate the rate constant at this temperature.

$$2N_2O(g) \rightarrow 2N_2(g) + O_2(g)$$

Solution

See Section 5.1

$$t_{\frac{1}{2}} = \frac{0.693}{k}$$

$$k = \frac{0.693}{t_{\frac{1}{2}}}$$

$$= \frac{0.693}{0.9 \text{ s}}$$

$$= 0.77 \text{ s}^{-1}$$

Example 5.7 Determining the rate constant of a first order reaction

$$2N_2O_5(g) \rightarrow 4NO_2(g) + O_2(g)$$

The rate of decomposition of N_2O_5 was studied at a particular temperature by monitoring the concentration of N_2O_5 every 100 s, and the following data was obtained.

Time (s)	$[N_2O_5]$ (mol dm^{-3})
0	0.10
100	0.06
200	0.036
300	0.023
400	0.013

    (a) Draw a time–concentration graph and deduce the half-life for the decomposition.

    (b) Decide whether the decomposition of N_2O_5 is first order or not.

    (c) Calculate the value of the rate constant for the decomposition of N_2O_5.

Solution

(a)

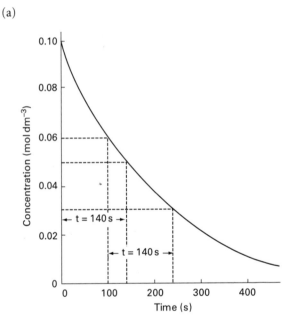

Consider the time taken for the concentration to half from 0.10 to 0.05 mol dm^{-3}. The graph shows that this takes 140 seconds. Any other halving period should take the same time – look at the change from 0.06 to 0.03 mol dm^{-3} for example. This means that the half-life of this reaction is 140 seconds.

(b) Since $t_{\frac{1}{2}}$ is constant, the reaction is first order.

(c)

$$k = \frac{0.693}{t_{\frac{1}{2}}}$$

$$k = \frac{0.693}{140 \text{ s}}$$

$$= 4.95 \times 10^{-3} \text{ s}^{-1}$$

5.4 Reaction mechanisms

Example 5.8 Proposing reaction mechanisms

For the reaction

$$NO_2(g) + CO(g) \rightarrow NO(g) + CO_2(g)$$

the rate equation is

$$\text{rate} = k \, [NO_2]^2$$

Propose a reaction mechanism.

Hint: Note that the [CO] term does not appear in the rate equation and the order of reaction with respect to NO_2 is 2. The reaction may take place in more than one step and in the rate-determining step (slow step) two NO_2 molecules are involved.

Solution

The following mechanism can be proposed.

$$2NO_2 \rightarrow NO + NO_3 \quad \text{slow}$$

$$NO_3 + CO \rightarrow NO_2 + CO_2 \quad \text{fast}$$

5.5 Practice questions

1. State the order of the reaction with respect to each reactant and the overall order of each of the reactions given below.

(a) $$A(g) + 2B(g) \rightarrow C(g) + D(g)$$

$$\text{Rate} = k[A][B]$$

(b) $$AX(g) \rightarrow B(g) + C(g)$$

$$\text{Rate} = k[AX]^2$$

(c) $$2H_2(g) + 2NO(g) \rightarrow 2H_2O(g) + N_2(g)$$

$$\text{Rate} = k[H_2][NO]^2$$

2. The equation for the reaction between iodide ions and perdisulphate ions is

$$2I^-(aq) + S_2O_8^{2-}(aq) \rightarrow I_2(aq) + 2SO_4^{2-}(aq)$$

If the rate of formation of I_2 is 1.0×10^{-7} mol dm^{-3} s^{-1}, what is

(a) the rate of consumption of I^-?

(b) the rate of consumption of $S_2O_8^{2-}$?

(c) the rate of formation of SO_4^{2-}?

3. The equation for the reaction between NO and Cl_2 gases is

$$2NO(g) + Cl_2(g) \rightarrow 2NOCl(g)$$

The rate equation for the reaction is

$$rate = k\,[NO]^2\,[Cl_2]$$

How would the reaction rate be altered if

(a) $[Cl_2]$ was kept constant and $[NO]$ doubled?

(b) $[NO]$ was kept constant and $[Cl_2]$ doubled?

(c) $[NO]$ was kept constant and $[Cl_2]$ halved?

(d) both $[NO]$ and $[Cl_2]$ were doubled?

4. For the reaction

$$X + Y \rightarrow Z$$

the reaction rate remains the same when the concentration of X is doubled, and the reaction rate is increased four times when the concentration of Y is doubled. Write the rate equation for the reaction.

5. Ethanal, CH_3CHO, decomposes to methane and carbon monoxide.

$$CH_3CHO(g) \rightarrow CH_4(g) + CO(g)$$

The initial rate of decomposition of ethanal was studied by taking different initial concentrations of ethanal at a particular temperature and the results are given below.

Experiment	Initial $[CH_3CHO]$ (mol dm^{-3})	Initial rate of decomposition (mol dm^{-3} s^{-1})
1	0.05	0.005
2	0.10	0.02
3	0.20	0.08

(a) Deduce the rate equation

(b) Calculate the value of the rate constant and deduce its unit.

(c) Calculate the initial rate of decomposition for an initial concentration of 0.5 mol dm^{-3} of ethanal.

6. For the reaction,

$$XO_3^-(aq) + 5X^-(aq) + 6H^+(aq) \rightarrow 3X_2(aq) + 3H_2O(I)$$

the following experimental data was obtained.

Experiment	Initial $[XO_3^-]$ (mol dm^{-3})	Initial $[X^-]$ (mol dm^{-3})	Initial $[H^+]$ (mol dm^{-3})	Initial rate of formation of X_2 (mol dm^{-3} s^{-1})
1	0.1	0.5	0.6	2.4×10^{-4}
2	0.1	0.5	1.2	9.6×10^{-4}
3	0.2	0.5	1.2	1.9×10^{-3}
4	0.1	0.25	1.2	4.8×10^{-4}

(a) Deduce the order of reaction with respect to each of the reactants.

(b) Write the rate equation.

(c) Calculate the value of the rate constant for the reaction and deduce its unit.

(d) What is the initial rate of the reaction when a mixture containing 0.1 mol dm^{-3} each of XO_3^-, X^- and H^+ is prepared?

7. For the reaction

$$2N_2O(g) \rightarrow 2N_2(g) + O_2(g)$$

the rate equation is,

$$\text{rate} = k\,[N_2O]$$

Propose a reaction mechanism.

8. For the reaction between nitrogen(II) oxide and hydrogen to form nitrogen and water, the overall equation is

$$2NO(g) + 2H_2(g) \rightarrow N_2(g) + 2H_2O(g)$$

Deduce the rate equation for the reaction, given the following reaction mechanism.

$$2NO(g) + H_2(g) \rightarrow N_2O(g) + H_2O(g) \quad \text{slow}$$

$$N_2O(g) + H_2(g) \rightarrow N_2(g) + H_2O(g) \qquad \text{fast}$$

Oxidation–Reduction, electrolysis, electrochemical cells

One-minute overview

In this chapter we will examine important concepts relating to oxidation–reduction reactions (in terms of electron transfer and change in oxidation number), electrolysis and electrochemical cells. The rules for obtaining oxidation numbers of atoms in substances are described. Diagrams for a typical electrolytic cell and an electrochemical cell are illustrated.

Sample examples include:
■ half equations for oxidation and reduction reactions
■ overall ionic equations for redox reactions
■ assigning oxidation numbers to atoms in substances
■ deciding whether a given reaction is redox or not
■ calculating quantities of substances produced during electrolysis
■ determining the feasibility of redox reactions using electrode potentials
■ calculating standard cell potentials.

6.1 Redox reactions: in terms of electron transfer

Oxidation is electron loss. A typical half equation for oxidation is

$$Mg \rightarrow Mg^{2+} + 2e^-$$

$\frac{1}{2}O_2$ represents 0.5 mole of oxygen molecules. This provides 1 mole of oxygen atoms, each of which accepts two electrons to form an oxide ion, O^{2-}.

Reduction is electron gain. A typical half equation for reduction is

$$\frac{1}{2}O_2 + 2e^- \rightarrow O^{2-}$$

A **redox** reaction is one in which oxidation and reduction take place simultaneously.

A typical overall equation for a redox reaction is

$$Mg + \frac{1}{2}O_2 \rightarrow MgO$$

An **oxidant (oxidising agent)** is an electron acceptor. Oxygen is the oxidant in the above example.

A **reductant (reducing agent)** is an electron donor. Magnesium is the reductant in the above example.

Note that:

- electrons appear on the right-hand side of the half equation for oxidation and left-hand side for reduction;

- to get an overall equation for a redox reaction, add the half equation for oxidation and the half equation for reduction, after equalising the number of electrons in the two half equations;

- electrons do not appear in an overall equation for a redox reaction as the electrons lost during oxidation are used up during reduction.

Conjugate oxidant–reductant pairs differ only in the number of electrons they carry.

A **conjugate oxidant–reductant pair** is an oxidant and its corresponding reduced form. Few examples of conjugate pairs are Mg^{2+} and Mg (in the above example); Fe^{3+} and Fe; and Cl_2 and Cl^- (in Example 6.1)

6.2 Writing half and overall ionic equations for redox reactions

Example 6.1 Writing half and overall ionic equations

Iron reacts with chlorine gas to form iron(III) chloride. Write half equations for the oxidation and reduction reactions, and hence an overall equation for the redox reaction.

Solution

Step 1: The formula of iron(III) chloride is $FeCl_3$; it consists of Fe^{3+} and Cl^- ions. An Fe atom donates three electrons and forms Fe^{3+}. A Cl_2 molecule consists of two Cl atoms, and each Cl atom accepts one electron to form a Cl^- ion. The half equations for oxidation and reduction are:

$$\text{oxidation} \quad Fe \rightarrow Fe^{3+} + 3e^- \qquad (1)$$

$$\text{reduction} \quad Cl_2 + 2e^- \rightarrow 2Cl^- \qquad (2)$$

Step 2: Equalise the numbers of electrons donated and accepted by multiplying equation (1) by 2 and equation (2) by 3. Add the two equations cancelling out the electrons from both sides.

$$2Fe + 3Cl_2 + \cancel{6e^-} \rightarrow 2Fe^{3+} + \cancel{6e^-} + 6Cl^-$$

This gives the overall ionic equation for the redox reaction,

$$2Fe + 3Cl_2 \rightarrow 2Fe^{3+} + 6Cl^- (\rightarrow 2FeCl_3)$$

Example 6.2

The manganate(VII) ion (MnO_4^-) oxidises Fe^{2+} to Fe^{3+} in the presence of an acid. MnO_4^- is reduced to Mn^{2+}. Write half equations for the oxidation and reduction reactions, and hence an overall equation for the redox reaction.

Solution

This is a more difficult example. To write the half equation for the reduction of MnO_4^-, follow *Steps 1–3* given below.

Step 1: The oxidant MnO_4^- ion is reduced to Mn^{2+}.

$$MnO_4^- \rightarrow Mn^{2+}$$

Step 2: Add $4H_2O$ to the right-hand side of the equation to balance the oxygen atoms in MnO_4^-, and then $8H^+$ (available from the acid) to the left-hand side to balance the hydrogen atoms in $4H_2O$.

$$MnO_4^- + 8H^+ \rightarrow Mn^{2+} + 4H_2O$$

Step 3: Add $5e^-$ to the left-hand side to balance the number of charges. Remember that an oxidant (MnO_4^- in this example) is an electron acceptor. This gives the half equation for the reduction of MnO_4^-.

$$MnO_4^- + 8H^+ + 5e^- \rightarrow Mn^{2+} + 4H_2O \text{ (reduction)}$$

Step 4: The half equation for the oxidation of Fe^{2+} to Fe^{3+} is

$$Fe^{2+} \rightarrow Fe^{3+} + e^- \text{ (oxidation)}$$

Step 5: Multiply the equation for the oxidation by 5 to balance for the number of electrons lost and gained, and then add to the equation for reduction to get the overall equation for the redox reaction (the electrons cancel out.)

$$MnO_4^- + 8H^+ + 5Fe^{2+} \rightarrow Mn^{2+} + 4H_2O + 5Fe^{3+}$$

6.3 Oxidation number

Oxidation number (ON) is the number of charges an atom 'would have' in a substance if electrons were transferred completely and is calculated according to a set of rules, which is given below.

1. All atoms in the elemental state are assigned an ON of zero. For example,

$$\overset{0}{H_2} \quad \overset{0}{Na} \quad \overset{0}{C} \quad \overset{0}{Al}$$

2. The oxidation number of hydrogen in all its compounds is +1, except in metal hydrides in which it is −1. For example,

$$\overset{+1}{H_2}O, \quad \overset{+1}{CH_4}, \quad \overset{+1}{H_2}SO_4, \quad Na\overset{+1}{H}CO_3, \quad Na\overset{-1}{H}$$

> In all these examples, the oxidation number of an atom is written above it.

3. The oxidation number of oxygen in all its compounds is −2, except in peroxides (in which it is −1) and in fluorine oxide, F_2O, (in which it is +2). For example,

$$H_2\overset{-2}{O}, \quad HN\overset{-2}{O}_3, \quad C\overset{-2}{O}Cl_2, \quad CaC\overset{-2}{O}_3, \quad H_2\overset{-1}{O}_2$$

4. The oxidation number of a simple ion equals the charge on the ion. For example,

$$\overset{+1}{Na^+} \quad \overset{+3}{Fe^{3+}} \quad \overset{-1}{Cl^-} \quad \overset{+2}{Ca^{2+}} \quad \overset{-2}{S^{2-}}$$

5. In a covalent molecule, the more electronegative atom assumes a negative oxidation number and the less electronegative atom has a positive number. For example, the oxidation number of C in CH_4 is −4 since the oxidation number of H is +1 (Rule 2) and C is more electronegative than H. In CCl_4, the oxidation number of C is +4, Cl being more electronegative than C.

6. The sum of the oxidation numbers of all the atoms in a molecule equals zero, and that of an ion equals the charge on the ion. Some examples are given below.

$$CH_4: \overset{-4}{C}(\overset{+1}{H})_4 \quad CCl_4: \overset{+4}{C}(\overset{-1}{Cl})_4 \quad SCl_2: \overset{+2}{S}(\overset{-1}{Cl})_2 \quad NH_3: \overset{-3}{N}(\overset{+1}{H})_3$$

$$NO_3^-: \overset{+5}{N}(\overset{-2}{O})_3^- \quad NH_4^+: \overset{-3}{N}(\overset{+1}{H})_4^+ \quad SO_4^{2-}: \overset{+6}{S}(\overset{-2}{O})_4^{2-}$$

$$Cr_2O_7^{2-}: \overset{+6}{Cr_2}(\overset{-2}{O})_7^{2-}$$

6.4 Assigning oxidation numbers

Example 6.3 Assigning oxidation numbers

Assign oxidation numbers to all the atoms in the following compounds.

(a) $FeCl_2$ (b) Fe_2O_3 (c) $CHCl_3$ (d) $NaNO_3$ (e) NH_4Br

Solution

(a) In $FeCl_2$, the ON of Fe is $+2$ and that of Cl is -1 and the total charge of all the species in $FeCl_2$ equals to $[(+2) + (2 \times -1)] = 0$.

In a similar way, the oxidation numbers of atoms in the other compounds have been worked out below.

(b) $\overset{+3 \ -2}{Fe_2O_3}$ (c) $\overset{+2+1 \ -1}{CHCl_3}$ (d) $\overset{+1 \ +5 \ -2}{NaNO_3}$ (e) $\overset{-3 +1 \ -1}{NH_4Br}$

Example 6.4

Deduce the oxidation numbers of all the atoms in the HSO_4^- ion.

Solution

The ON of H is $+1$ and that of O is -2. The sum of the oxidation numbers of H, S and O should be equal to -1, which is the charge on the ion. So the ON of S must be $+6$. That is, the sum of all the oxidation numbers in HSO_4^- equals $[(+1) + (+6) + (4 \times -2)] = -1$.

$$\overset{+1 +6 -2}{HSO_4^-}$$

Example 6.5 Assigning oxidation numbers to atoms in ions

Assign oxidation numbers to all the atoms in the following ions.

(a) NO_2^- (b) SO_3^{2-} (c) MnO_4^- (e) CO_3^{2-}

Solution

(a) In NO_2^-, the oxidation number of N is $+3$ and that of O is -2.

(b) In SO_3^{2-}, the oxidation number of S is $+4$ and that of O is -2.

(c) In MnO_4^-, the oxidation number of Mn is $+7$ and that of O is -2.

(d) In CO_3^{2-}, the oxidation number of C is $+4$ and that of O is -2.

Example 6.6 Variable oxidation numbers for the same element

For example, chlorine exhibits oxidation numbers -1 to $+7$ in its compounds. Work out the oxidation number of Cl in the following substances. (Oxidation numbers: H = +1, O = −2, Cu = +2, Na = +1, K = +1, F = −1)

(a) $HClO_2$ (b) $CuCl_2$ (c) $NaClO_3$ (d) $KClO_4$ (e) $NaClO$

(f) ClF (g) Cl_2O

Solution

(a) $HClO_2$: the sum of the oxidation numbers of H (+1) and two O atoms (2 × −2) is −3, so Cl is +3.

(b) $CuCl_2$: the oxidation number of Cu is +2, so Cl is −1.

(c) $NaClO_3$: the sum of the oxidation numbers of Na (+1) and three O atoms (3 × −2) is −5, so Cl is +5.

(d) $KClO_4$: the sum of the oxidation numbers of K (+1) and four O atoms (4 × −2) is −7, so Cl is +7.

(e) $NaClO$: the sum of the oxidation numbers of Na (+1) and O (−2) is −1, so Cl is +1.

(f) ClF: the oxidation number of F is −1, so Cl is +1.

(g) Cl_2O: the oxidation number of O is −2, so Cl is +1.

Example 6.7 Using oxidation numbers to find the charge on a species

Use the given oxidation numbers of the elements to deduce whether the following species are ions or molecules.

(H = +1, O = −2, Mn = +4, I = +5, P = +5, N = +5, S = +4)

(a) IO_3 (b) MnO_2 (c) PO_4 (d) HNO_3 (e) HSO_3

Solution

(a) IO_3: the sum of the oxidation numbers of I (+5) and three O atoms (3 × −2 = −6) equals −1. So it is an ion of formula IO_3^-.

(b) MnO_2: the sum of the oxidation numbers of Mn (+4) and two O atoms ($2 \times -2 = -4$) is 0, so MnO_2 is a molecule.

(c) PO_4: the sum of the oxidation numbers of P (+5) and four O atoms ($4 \times -2 = -8$) equals –3, so it is an ion of formula PO_4^{3-}.

(d) HNO_3: the sum of the oxidation numbers of H (+1), N (+5) and three O atoms ($3 \times -2 = -6$) equals 0, so HNO_3 is a molecule.

(e) HSO_3: the sum of the oxidation numbers of H (+1), S (+4) and three O atoms ($3 \times -2 = -6$) equals –1, so it is an ion of formula HSO_3^-.

6.5 Redox reactions: in terms of change in oxidation number

Oxidation involves an increase in oxidation number, and **reduction** involves a decrease in oxidation number during a redox reaction.

An **oxidant** is a species which decreases in oxidation number, and a **reductant** is a species which increases in oxidation number during a redox reaction.

Disproportionation is a redox reaction in which one element undergoes oxidation and reduction simultaneously. See Example 6.9.

Example 6.8 Identifying redox reactions using oxidation numbers

Work out the oxidation numbers of all the atoms in the reactants and products and decide which of the following are redox reactions. Name the oxidant and reductant in each of the redox reactions.

(a) $\qquad Mg + 2HCl \quad \rightarrow \quad MgCl_2 + H_2$

(b) $\qquad MgO + 2HCl \rightarrow MgCl_2 + H_2O$

Solution

(a)
$$\overset{0}{Mg} + 2\overset{+1\,-1}{HCl} \rightarrow \overset{+2\,-1}{MgCl_2} + \overset{0}{H_2}$$

It is a redox reaction. The ON of Mg is increased from 0 to +2, so Mg is oxidised. Mg is the reductant. The ON of H (in HCl) is decreased from +1 to 0 (in H_2), so H in HCl is reduced. HCl is the oxidant.

(b)
$$\overset{+2\,-2}{MgO} + 2\overset{+1\,-1}{HCl} \rightarrow \overset{+2\,-1}{MgCl_2} + \overset{+1\,-2}{H_2O}$$

There is no change in the oxidation numbers of the atoms and it is therefore not a redox reaction (it is a neutralisation, or acid–base, reaction).

Example 6.9 Identifying disproportionation reactions

Assign oxidation numbers to all atoms in the reactants and products in the reaction given below. Name the oxidant and the reductant.

$$3NO_2(g) + H_2O(l) \rightarrow 2HNO_3(aq) + NO(g)$$

Solution

$$3\overset{+4\,-2}{NO_2}(g) + \overset{+1\,-2}{H_2O}(l) \rightarrow 2\overset{+1\,+5\,-2}{HNO_3}(aq) + \overset{+2\,-2}{NO}(g)$$

The oxidation number of N, +4 in NO_2, is decreased to +2 in NO and also increased to +5 in HNO_3. NO_2 undergoes both reduction and oxidation. NO_2 is the oxidant as well as the reductant. This is an example of a disproportionation reaction.

> Note that you can ignore the stoichiometric numbers in an equation when using oxidation numbers to identify redox.

6.6 Electrolysis

Electrolysis is a chemical reaction that takes place during the passage of an electric current through an electrolyte.

An **electrolytic cell** is a set of apparatus consisting of two electrodes placed in an electrolyte in which an electric current is used to bring about chemical change.

Figure 6.1
A typical
electrolytic cell
▶

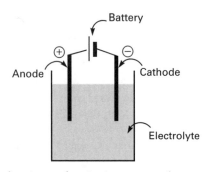

An **electrolyte** is a molten ionic compound or an aqueous ionic solution which conducts electricity.

The **electrodes** are graphite or metal rods used to carry electricity to the electrolyte.

The **anode** is the positive electrode which takes electrons from the negative ions attracted to it. It is the electrode at which oxidation occurs (the negative ions lose electrons).

The **cathode** is the negative electrode which gives electrons to the positive ions. It is the electrode at which reduction takes place (the positive ions accept electrons).

When electricity is passed through a **simple ionic compound** in the molten state, the metal ions accept electrons from the cathode and are reduced to metal atoms. At the same time the negative ions donate electrons to the anode (which move through the external wire toward the cathode) and are oxidised to non-metal atoms. For example, when molten calcium bromide is electrolysed, calcium is produced at the cathode and bromine at the anode.

$$\text{At the cathode} \quad Ca^{2+} + 2e^- \rightarrow Ca$$

$$\text{At the anode} \quad 2Br^- \rightarrow Br_2 + 2e^-$$

When an **aqueous ionic solution** is electrolysed, the products depend on the position of the ions in the electrochemical series (see Section 6.7), the concentration of the ions and the type of electrode used. For example, when copper(II) sulphate solution is electrolysed between graphite electrodes, copper is deposited on the cathode and oxygen is evolved at the anode. The solution contains Cu^{2+} and SO_4^{2-} ions from $CuSO_4$, and H^+ and OH^- ions from H_2O. During electrolysis the cations Cu^{2+} and H^+ move toward the cathode, but Cu^{2+} is discharged in preference to H^+ (Cu is below H in the electrochemical series). SO_4^{2-} and OH^- ions move toward the anode, but OH^- is discharged in preference to SO_4^{2-}.

$$\text{At the cathode} \quad Cu^{2+} + 2e^- \rightarrow Cu$$

$$\text{At the anode} \quad 4OH^- \rightarrow 2H_2O + O_2 + 4e^-$$

If the electrolysis is done using copper electrodes, copper is deposited on the cathode, as in the previous case. But at the anode, Cu atoms are oxidised providing electrons. These electrons move from the anode to the cathode through the external wire while the Cu^{2+} ions move through the electrolyte towards the cathode.

$$\text{At the cathode} \quad Cu^{2+} + 2e^- \rightarrow Cu$$

$$\text{At the anode} \quad Cu \rightarrow Cu^{2+} + 2e^-$$

Faraday's law of electrolysis the number of moles of a product formed during electrolysis is stoichiometrically equivalent to the amount of electricity (number of moles of electrons) supplied.

The unit of electric charge is **coulomb** (C) and one coulomb is the quantity of electricity passed when a current of one ampere is passed for one second.

$$\begin{array}{c} \text{Quantity of electricity} \\ \text{in coulombs} \end{array} = \begin{array}{c} \text{current in amperes} \\ \times \text{ time in seconds} \end{array}$$

or $$Q = I\,t$$

Faraday's constant is the charge carried by 1 mole of electrons and its value is 96 500 coulombs.

$$1 \text{ mole of electrons} \equiv 6.022 \times 10^{23}\, e^- \equiv 96\,500 \text{ C} \equiv 1 \text{ F}$$

From the quantity of electricity used for the electrolysis, the number of moles of electrons used can be calculated, which in turn can be used to calculate the mass of a product.

Example 6.10 Calculating the molar amount and mass of an electrolytic product

A solution of copper(II) sulphate is electrolysed using an electric current of 5.0 A for 10 min. Calculate:

(a) the quantity of electricity used,

(b) the corresponding number of moles of electrons,

(c) the number of moles of copper deposited on the cathode,

(d) the mass of copper produced.

(A_R: Cu = 63.6)

Solution

(a)
$$Q = I\,t$$
$$= 5.0\,\text{A} \times (10 \times 60)\text{s}$$
$$= 3\,000\,\text{C}$$

(b)
$$96\,500\,\text{C} \equiv 1 \text{ mole of e}^-$$
$$\text{Moles of e}^- \text{ in } 3\,000\,\text{C} = \frac{3\,000\,\text{C}}{96\,500\,\text{C mol}^{-1}}$$
$$= 0.031\,\text{mol}$$

(c)
$$Cu^{2+} + 2e^- \rightarrow Cu$$
$$1\text{ mol}\quad 2\text{ mol}\quad 1\text{ mol}$$
$$1 \text{ mole of Cu} \equiv 2 \text{ mole of e}^-$$
$$\text{Moles of Cu} = \frac{0.031}{2} = 0.0155\,\text{mol}$$

(d)   Mass of Cu = moles × molar mass

$$= 0.0155 \text{ mol} \times 63.6 \text{ g mol}^{-1}$$

$$= 0.99 \text{ g}$$

Example 6.11 Calculating the time to produce a given amount of substance

Calculate the time, in minutes, needed to produce 1.21 g of magnesium by the electrolysis of molten magnesium chloride using a current of 10.0 A. (A_R: Mg = 24.0)

Solution

$$\text{Moles of Mg} = \frac{1.21 \text{ g}}{24.0 \text{ g mol}^{-1}} = 0.05 \text{ mol}$$

$$Mg^{2+} + 2e^- \rightarrow Mg$$
$$1 \text{ mol} \quad 2 \text{ mol} \quad 1 \text{ mol}$$

Moles of e⁻ needed to produce 0.05 mol of Mg
$$= 2 \times 0.05 \text{ mol} = 0.10 \text{ mol}$$

$$1 \text{ mol } e^- \equiv 96\,500 \text{ C}$$

Therefore, 0.1 mol e⁻ ≡ 9 650 C

$$Q = I\,t$$

$$t = \frac{9\,650 \text{ C}}{10 \text{ A}} = 965 \text{ s}$$

$$= 16.1 \text{ min}$$

Example 6.12 Calculating the mass or volume of a gas produced during electrolysis

An aqueous solution of sodium chloride is electrolysed, under suitable conditions, to produce hydrogen at the cathode and chlorine at the anode using a 2.0 A current for half an hour. Calculate

(a) the mass of hydrogen,

(b) the volume of chlorine at 298 K and 1 atm pressure.

[A_R: H = 1.0. Molar volume of a gas at 298 K and 1 atm. = 24.0 dm^3]

Solution

(a)
$$Q = I\,t$$
$$= 2.0\ \text{A} \times (30 \times 60)\text{s}$$
$$= 3600\ \text{C}$$

$$\text{Moles of e}^- = \frac{3600}{96500} = 0.037\ \text{mol}$$

$$2\text{H}^+ + 2\text{e}^- \rightarrow \text{H}_2$$
1 mol   2 mol   1 mol

$$\text{Moles of H}_2 = \frac{0.037}{2} = 0.0185\ \text{mol}$$

$$\text{Mass of H}_2 = 0.0185\ \text{mol} \times 2\ \text{g mol}^{-1} = 0.037\ \text{g}$$

(b) When 1 mole of H$_2$ is made at the cathode, 1 mole of chlorine is made at the anode.

$$2\text{Cl}^- \rightarrow \text{Cl}_2 + 2\text{e}^-$$
1 mol   1 mol   2 mol

$$\text{Moles of Cl}_2 = \frac{0.037}{2} = 0.0185\ \text{mol}$$

$$\text{Volume of Cl}_2 = 0.0185\ \text{mol} \times 24\ \text{dm}^3\ \text{mol}^{-1}$$
$$= 0.444\ \text{dm}^3 \text{ at 298 K and 1 atm.}$$

6.7 Electrochemical cells

An **electrochemical cell** is a set of apparatus in which a redox reaction is carried out in order to produce an electric current.

An electrochemical cell consists of two **half cells** or **electrodes**. One half cell contains an oxidant and the other a reductant, both under standard conditions (explained below). They are coupled through a voltmeter and a salt bridge to form a **standard cell**.

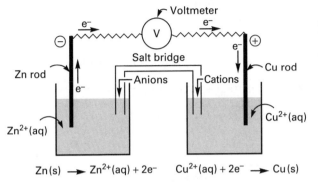

Figure 6.2
An electrochemical cell

$$Zn(s) \longrightarrow Zn^{2+}(aq) + 2e^- \qquad Cu^{2+}(aq) + 2e^- \longrightarrow Cu(s)$$

The shorthand notation for the above electrochemical cell is

$$Zn(s) \mid Zn^{2+}(aq) \parallel Cu^{2+}(aq) \mid Cu(s)$$

There are **three types of half cells.**

- A metal rod placed in its metal ion solution – for example $Zn(s) \mid Zn^{2+}(aq)$.

- A gas passed through its oxidised or reduced form of ions in solution, both in contact with an inert electrode. For example $H_2(g)$, $H^+(aq) \mid Pt(s)$. In this case, H_2 gas is bubbled over a platinum electrode in an acidic solution, such as $H_2SO_4(aq)$.

- Both the oxidant and its reduced species, or reductant and its oxidised species, in solution in contact with an inert electrode – for example $Fe^{2+}(aq)$, $Fe^{3+}(aq) \mid Pt(s)$.

In a *standard half cell*, the concentrations of all ions must be 1 mol dm^{-3} and any gas must be at a pressure of 1 atm, all at a temperature of 298 K.

The **standard electrode potential,** $E^{\ominus}$ of a standard hydrogen electrode is assigned a value of 0.0 volts (V). The standard electrode potential of any other electrode is the potential difference between that electrode and a hydrogen electrode, both electrodes under standard conditions. Some standard electrode potentials are listed in Table 6.1.

$E^{\ominus}$ values are generally quoted for reduction reactions. For oxidation reactions the sign of $E^{\ominus}$ is reversed. Note the examples given below.

$$Cu^{2+}(aq) + 2e^- \rightarrow Cu(s) \ E^{\ominus}_{red} = +0.34 \text{ V}$$

$$2H^+(aq) + 2e^- \rightarrow H_2(g) \ E^{\ominus}_{red} = 0.00 \text{ V}$$

$$Cu(s) \rightarrow Cu^{2+}(aq) + 2e^- \ E^{\ominus}_{ox} = -0.34 \text{ V}$$

$$H_2(g) \rightarrow 2H^+(aq) + 2e^- \ E^{\ominus}_{ox} = 0.00 \text{ V}$$

Table 6.1
Electrode
potentials
▶

Reaction		$E^{\ominus}$(V)
$F_2(g) + 2e^-$	$\rightarrow 2F^-(aq)$	+2.87
$H_2O_2(aq) + 2H^+(aq) + 2e^-$	$\rightarrow 2H_2O(l)$	+1.77
$MnO_4^-(aq) + 8H^+ + 5e^-$	$\rightarrow Mn^{2+}(aq) + 4H_2O(l)$	+1.51
$Cl_2(g) + 2e^-$	$\rightarrow 2Cl^-(aq)$	+1.36
$Cr_2O_7^{2-}(aq) + 14H^+ + 6e^-$	$\rightarrow 2Cr^{3+}(aq) + 7H_2O(l)$	+1.33
$Br_2(g) + 2e^-$	$\rightarrow 2Br^-(aq)$	+1.09
$Ag^+(aq) + e^-$	$\rightarrow Ag(s)$	+0.80
$Fe^{3+}(aq) + e^-$	$\rightarrow Fe^{2+}(aq)$	+0.77
$2H^+(aq) + O_2(g) + 2e^-$	$\rightarrow H_2O_2(aq)$	+0.68
$I_2(aq) + 2e^-$	$\rightarrow 2I^-(aq)$	+0.54
$Cu^+(aq) + e^-$	$\rightarrow Cu(s)$	+0.52
$O_2(g) + 2H_2O(l) + 4e^-$	$\rightarrow 4OH^-(aq)$	+0.40
$Cu^{2+}(aq) + 2e^-$	$\rightarrow Cu(s)$	+0.34
$Cu^{2+}(aq) + e^-$	$\rightarrow Cu^+(aq)$	+0.15
$2H^+(aq) + 2e^-$	$\rightarrow H_2(g)$	0.00
$Pb^{2+}(aq) + 2e^-$	$\rightarrow Pb(s)$	−0.13
$Sn^{2+}(aq) + 2e^-$	$\rightarrow Sn(s)$	−0.14
$Ni^{2+}(aq) + 2e^-$	$\rightarrow Ni(s)$	−0.25
$Fe^{2+}(aq) + 2e^-$	$\rightarrow Fe(s)$	−0.44
$Zn^{2+}(aq) + 2e^-$	$\rightarrow Zn(s)$	−0.76
$Al^{3+}(aq) + 3e^-$	$\rightarrow Al(s)$	−1.66
$Mg^{2+}(aq) + 2e^-$	$\rightarrow Mg(s)$	−2.37
$Na^+(aq) + e^-$	$\rightarrow Na(s)$	−2.71
$Ca^{2+}(aq) + 2e^-$	$\rightarrow Ca(s)$	−2.87

The **standard cell potential** ($E^{\ominus}_{\text{cell}}$) of an electrochemical cell can be obtained by adding the standard electrode potentials for the oxidation and reduction reactions that take place in the cell (see Example 6.13). If the standard cell potential is positive, the redox reaction is spontaneous. If it is negative, the redox reaction is not spontaneous, but is spontaneous in the opposite direction.

Example 6.13

$$Zn^{2+}(aq) + 2e^- \rightarrow Zn(s) \qquad E^{\ominus} = -0.76 \text{ V}$$

$$Cu^{2+}(aq) + 2e^- \rightarrow Cu(s) \qquad E^{\ominus} = +0.34 \text{ V}$$

A $Zn(s) \mid Zn^{2+}(aq)$ half cell is coupled with a $Cu(s) \mid Cu^{2+}(aq)$ half cell. Use the standard electrode potentials to answer the following questions.

(a) Write half equations for the reactions that take place in the half cells.

(b) Write the equation for the overall reaction.

(c) What is the standard cell potential?

Solution

In this case, oxidation takes place at the Zn electrode since the oxidation potential of zinc is more positive ($+0.76$ V) than the reduction potential of Cu electrode ($+0.34$ V).

(a) $\qquad Zn(s) \rightarrow Zn^{2+}(aq) + 2e^- \; E^{\ominus} = +0.76 \text{ V}$

$\qquad\qquad Cu^{2+}(aq) + 2e^- \rightarrow Cu(s) \; E^{\ominus} = +0.34 \text{ V}$

(b) $\qquad Zn(s) + Cu^{2+}(aq) \rightarrow Zn^{2+}(aq) + Cu(s)$

(c) $\qquad$ The cell potential $= +0.76 + (+0.34) \text{ V} = +1.10 \text{ V}$

> **Hint:** Write one equation for oxidation and one for reduction in such a way that the sum of the oxidation and reduction potentials is *positive*, which indicates that the reaction takes place spontaneously.

Example 6.14 Would you expect a reaction if: (Use the $E^{\ominus}$ values in Table 6.1)

(a) Mg metal is added to a 1 mol dm^{-3} solution of HCl?

(b) Cu metal is added to a 1 mol dm^{-3} solution of HCl?

(c) $Fe^{3+}(aq)$ is added to $I^-(aq)$?

(d) I_2 (aq) is added to KCl (aq)?

Solution

(a) When Mg is added to 1 mol dm^{-3} HCl, Mg is oxidised (higher positive $E^{\ominus}_{ox}$ value) and H$^+$ is reduced. The partial and overall equations and the $E^{\ominus}_{cell}$ are

$$Mg(s) \rightarrow Mg^{2+}(aq) + 2e^-\ \ E^{\ominus} = +2.37\ V$$

$$2H^+(aq) + 2e^- \rightarrow H_2(g)\ \ E^{\ominus} = 0.00\ V$$

$$Mg(s) + 2H^+ (aq) \rightarrow Mg^{2+}(aq) + H_2(g),\ E^{\ominus}_{cell} = +2.37\ V$$

Since $E^{\ominus}_{cell}$ is positive, the reaction takes place spontaneously as written above.

(b) $$Cu(s) \rightarrow Cu^{2+}(aq) + 2e^-\ \ E^{\ominus} = -0.34\ V$$

$$2H^+(aq) + 2e^- \rightarrow H_2(g)\ \ E^{\ominus} = 0.00\ V$$

$$Cu(s) + 2H^+(aq) \rightarrow Cu^{2+}(aq) + H_2(g),\ E^{\ominus}_{cell} = -0.34\ V$$

Since $E^{\ominus}_{cell}$ is negative, the above reaction does not take place spontaneously.

> Multiplying a half equation by 2, say, to balance the number of electrons in the other half equation does *not* alter the $E^{\ominus}$ value.

(c) $$2Fe^{3+}(aq) + 2e^- \rightarrow 2Fe^{2+}(aq)\ \ E^{\ominus} = +0.77\ V$$

$$2I^-(aq) \rightarrow I_2(aq) + 2e^-\ \ E^{\ominus} = -0.54\ V$$

$$2Fe^{3+}(aq) + 2I^-(aq) \rightarrow 2Fe^{2+}(aq) + I_2(aq),\ E^{\ominus}_{cell} = +0.23\ V$$

Since $E^{\ominus}_{cell}$ is positive, the above reaction takes place spontaneously. That is, Fe^{3+} oxidises I$^-$ to I$_2$ spontaneously.

(d) $$I_2(aq) + 2e^- \rightarrow 2I^-(aq)\ \ E^{\ominus} = +0.54\ V$$

$$2Cl^-(aq) \rightarrow Cl_2(g) + 2e^-\ \ E^{\ominus} = -1.36\ V$$

$$I_2(aq) + 2Cl^-(aq) \rightarrow 2I^-(aq) + Cl_2(g),\ E^{\ominus}_{cell} = -0.82\ V$$

Since $E^{\ominus}_{cell}$ is negative, the above reaction does not take place spontaneously. I$_2$ cannot oxidise Cl$^-$ to Cl$_2$.

Example 6.15 Arranging oxidants/reductants in order of reactivity

Arrange the following half cell reactions in decreasing order of their $E^\ominus$ values, indicating the order of increasing oxidising strength and increasing reducing strength. Name the strongest and weakest oxidants and the strongest and weakest reductants.

$$Fe^{2+}(aq) + 2e^- \rightarrow Fe(s) \qquad E^\ominus = -0.44 \text{ V}$$

$$Cu^{2+}(aq) + 2e^- \rightarrow Cu(s) \qquad E^\ominus = +0.34 \text{ V}$$

$$Zn^{2+}(aq) + 2e^- \rightarrow Zn(s) \qquad E^\ominus = -0.76 \text{ V}$$

$$2H^+(aq) + 2e^- \rightarrow H_2(g) \qquad E^\ominus = 0.0 \text{ V}$$

Solution

$Cu^{2+}(aq) + 2e^- \rightarrow Cu(s)$	$E^\ominus = +0.34$ V
increasing $\quad 2H^+(aq) + 2e^- \rightarrow H_2(g)$ $\quad$ increasing	$E^\ominus = 0.00$ V
oxidising $\qquad\qquad\qquad\qquad\qquad\qquad\quad$ reducing	
strength $\quad Fe^{2+}(aq) + 2e^- \rightarrow Fe(s) \quad$ strength	$E^\ominus = -0.44$ V
$Zn^{2+}(aq) + 2e^- \rightarrow Zn(s)$	$E^\ominus = -0.76$ V

Cu^{2+} is the strongest oxidant and Zn^{2+} the weakest oxidant. Zn is the strongest reductant and Cu the weakest reductant. (The stronger the oxidant the weaker its conjugate reductant and vice versa.)

Example 6.16 Predicting reactions using electrode potentials

Predict what happens when

(a) Cu metal is placed in $AgNO_3$ solution

(b) Ag metal is placed in $MgCl_2$ solution

(c) Cl_2 gas is passed through KBr solution

(d) Fe^{2+} ($FeSO_4$ solution) is added to acidified H_2O_2 solution.

Solution

(a)
$$Cu \rightarrow Cu^{2+} + 2e^- \quad E^{\ominus} = -0.34 \text{ V}$$

$$2Ag^+ + 2e^- \rightarrow 2Ag \quad E^{\ominus} = +0.80 \text{ V}$$

$$Cu + 2Ag^+ \rightarrow Cu^{2+} + 2Ag, \quad E^{\ominus}_{cell} = +0.46 \text{ V}$$

Since $E^{\ominus}_{cell}$ is positive, the reaction takes place spontaneously. Cu displaces Ag from $AgNO_3$ solution.

(b)
$$2Ag \rightarrow 2Ag^+ + 2e^- \quad E^{\ominus} = -0.80 \text{ V}$$

$$Mg^{2+} + 2e^- \rightarrow Mg \quad E^{\ominus} = -2.37 \text{ V}$$

$$2Ag + Mg^{2+} \rightarrow 2Ag^+ + Mg, \quad E^{\ominus}_{cell} = -3.17 \text{ V}$$

Since $E^{\ominus}_{cell}$ for the overall reaction is negative, Ag does not react spontaneously with Mg^{2+} to form Ag^+ and Mg.

(c)
$$2Br^- \rightarrow Br_2 + 2e^- \quad E^{\ominus} = -1.09 \text{ V}$$

$$Cl_2 + 2e^- \rightarrow 2Cl^- \quad E^{\ominus} = +1.36 \text{ V}$$

$$Cl_2 + 2Br^- \rightarrow 2Cl^- + Br_2, \quad E^{\ominus}_{cell} = +0.27 \text{ V}$$

$E^{\ominus}_{cell}$ for the overall reaction is positive so Cl_2 displaces Br_2 from KBr spontaneously.

(d)
$$2Fe^{2+} \rightarrow 2Fe^{3+} + 2e^- \quad E^{\ominus} = -0.77 \text{ V}$$

$$H_2O_2 + 2H^+ + 2e^- \rightarrow 2H_2O \quad E^{\ominus} = +1.61 \text{ V}$$

$$H_2O_2 + 2H^+ + 2Fe^{2+} \rightarrow 2H_2O + 2Fe^{3+}, \quad E^{\ominus}_{cell} = +0.84 \text{ V}$$

$E^{\ominus}_{cell}$ is positive. H_2O_2 oxidises Fe^{2+} to Fe^{3+} spontaneously, while it gets reduced to H_2O.

Example 6.17 Writing electrochemical cell notation and calculating cell potential

(a) Sketch a labelled diagram of an electrochemical cell for the following cell reaction. Indicate on the diagram the positive and negative electrodes and the direction in which the electrons migrate through the external circuit. Write equations for the electrode reactions.

$$H_2O_2 + 2H^+ + 2Fe^{2+} \rightarrow 2H_2O + 2Fe^{3+}$$

(b) Write the cell notation.

(c) What is the standard cell potential of this cell?

Solution

(a)

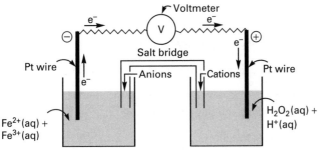

Equations for the electrode reactions:

at the negative electrode,

$$Fe^{2+}(aq) \rightarrow Fe^{3+}(aq) + e^-$$

at the positive electrode,

$$H_2O_2(aq) + 2H^+(aq) + 2e^- \rightarrow 2H_2O(l)$$

(b) Cell notation: $Fe^{2+}(aq), Fe^{3+}(aq) \mid Pt(s) \parallel H_2O_2(aq),$ $H^+(aq), H_2O(l) \mid Pt(s)$

(c) $2Fe^{2+}(aq) \rightarrow 2Fe^{3+}(aq) + 2e^-$ $E^{\ominus} = -0.77$ V

$H_2O_2(aq) + 2H^+(aq) + 2e^- \rightarrow 2H_2O(l)$ $E^{\ominus} = +1.61$ V

$2Fe^{2+}(aq) + H_2O_2(aq) + 2H^+(aq) \rightarrow 2Fe^{3+}(aq) + 2H_2O(l),$
$$E^{\ominus} = +0.84 \text{ V}$$

Standard cell potential = +0.84 V

6.8 Practice questions

1. Assign oxidation numbers to all the atoms in the following substances.

(a) P_2O_3, P_2O_5, CO, CO_2, SiO_2, BaO, Al_2O_3, Cr_2O_3

(b) $NaHCO_3$, $KMnO_4$, $CaCl_2$, AgCl, $CuBr_2$, $Ba(NO_3)_2$, $MgSO_4$, Na_2S

2. Assign oxidation numbers to all atoms in the following ions.

NO_3^-, SO_3^{2-}, CrO_4^{2-}, HSO_4^-, HPO_4^{2-}, HCO_3^-, OH^-, SiO_3^{2-}

3. Assign oxidation numbers to all the atoms in the following compounds of nitrogen.

KNO_2, $NaNO_3$, HNO_3, N_2O_5, N_2O, NO, N_2O_4, NO_2

4. Assign oxidation numbers to all the atoms in the following compounds of sulphur.

SO_2, SO_3, $NaHSO_3$, H_2SO_4, Na_2S, $ZnSO_4$, $KHSO_4$, H_2SO_3

5. Write ionic half equations for

(a) the oxidation of Ba to Ba^{2+}

(b) the oxidation of Br^- to Br_2

(c) the reduction of F_2 to F^-

(d) the oxidation of Cu^+ to Cu^{2+}

(e) the reduction of S to S^{2-}

6. Write ionic half equations for the oxidation and reduction reactions, and the overall ionic equations for

(a) the reaction between Na and Br_2 to form NaBr.

(b) the displacement of I_2 from I^- (aq) by Cl_2.

(c) the reaction between Cu metal and Ag^+ to form Cu^{2+} and Ag metal.

(d) the disproportionation of Cu^+ to Cu and Cu^{2+}.

7. Name the oxidant and reductant in these reactions.

(a) $Cl_2(g) + 2Fe^{2+}(aq) \rightarrow 2Cl^-(aq) + 2Fe^{3+}(aq)$

(b) $MnO_2(s) + 4HCl(aq) \rightarrow MnCl_2(aq) + Cl_2(g) + 2H_2O(l)$

(c) $2H_2S(aq) + SO_2(aq) \rightarrow 3S(s) + 2H_2O(l)$

(d) $\quad 3H_2(g) + N_2(g) \rightarrow 2NH_3(g)$

8. State which of the following are redox reactions.

(a) $Fe^{2+}(aq) + CO_3^{2-}(aq) \rightarrow FeCO_3(s)$

(b) $Cu(s) + 4HNO_3(aq) \rightarrow Cu(NO_3)_2(aq) + 2NO_2(g)$
$\qquad\qquad\qquad\qquad + 2H_2O(l)$

(c) $\quad H_2O(l) + SO_2(g) \rightarrow H_2SO_3(aq)$

(d) $\qquad 2NaHCO_3(s) \rightarrow Na_2CO_3(s) + CO_2(g)$
$\qquad\qquad\qquad\qquad + H_2O(g)$

(e) $Cl_2(g) + 2NaOH(aq) \rightarrow NaCl(aq) + NaClO(aq)$
$\qquad\qquad\qquad\qquad + H_2O(l)$

9. Calculate the amount of electricity passed when

(a) a 1.5 A current is passed for 2 hours

(b) a 4.0 A current is passed for 50 minutes

10. Calculate the number of moles of electrons transferred into an electrolyte when an electric current of 3.5 A is passed for 2 hours.

11. Calculate the number of moles of electrons needed to produce the following amounts of substances by electrolysis.

(a) 0.05 mole of Cu by the electrolysis of $CuSO_4$ solution

(b) 2 moles of Cl_2 gas by the electrolysis of molten NaCl

(c) 0.1 mole of O_2 gas by the electrolysis of dilute aqueous Na_2SO_4

(d) 0.5 mole of Ag by the electrolysis of aqueous $AgNO_3$

12. Calculate the mass of the metal produced at the cathode by the passage of a current of 12.0 A for an hour through molten $MgCl_2$. (A_R: Mg = 24.3)

13. Write equations for the reactions that take place at the electrodes when molten forms of the following substances are electrolysed.

    (a) Zinc chloride, $ZnCl_2$

    (b) Sodium bromide, NaBr

    (c) Aluminium oxide, Al_2O_3

14. A solution of copper(II) sulphate was electrolysed using a large block of impure copper as the anode and a thin sheet of pure copper as the cathode, and passing an electric current of 3.2 A for 6 hours. During electrolysis, copper dissolves from the anode and pure copper is deposited on the cathode. Calculate:

    (a) the quantity of electricity passed

    (b) the corresponding number of moles of electrons

    (c) the number of moles of copper deposited on the cathode

    (d) the mass of copper deposited. (A_R: Cu = 63.6)

15. Separate the reaction

    (a) $$Mg + Cl_2 \rightarrow MgCl_2$$

    into two ion–electron half equations.

    (b) Draw a labelled diagram of a cell to make use of the above reaction. Indicate the direction of the flow of electrons in the external circuit.

    (c) Use the electrode potential values given in Table 6.1 to find the standard cell potential.

16. (a) Use the electrode potential values given in Table 6.1 to calculate the standard cell potential in which the following overall reaction takes place.

$$Fe(s) + Cu^{2+}(aq) \rightarrow Fe^{2+}(aq) + Cu(s)$$

(b) Write the notation for the cell.

(c) Write ionic equations for the electrode reactions if the Cu (s) | Cu^{2+}(aq) electrode of the above cell is replaced by a Pb (s) | Pb^{2+}(aq) electrode. What is the new standard cell potential?

7 ▮ Acids and bases

One-minute overview

Two important concepts about acids and bases are examined, followed by a discussion of the acid constant, base constant, ionic product of water, strong and weak acids and bases, and buffer solutions.

Examples include calculations involving:
- ionic product of water
- acid and base constants
- pH of strong acids/bases
- pH of weak acids/bases
- buffer solutions.

7.1 Important terms, definitions and relationships

The two key concepts relating to acids and bases are the Lewis concept and the Bronsted–Lowry concept.

The Lewis concept of acids and bases

A **Lewis acid** is an electron pair acceptor and a **Lewis base** is an electron pair donor. H^+, Cu^{2+} and BF_3 are examples of Lewis acids, and H_2O, NH_3 and OH^-, are examples of Lewis bases. In the reaction

$$\underset{\text{acid}}{BH_3} + \underset{\text{base}}{NH_3} \rightleftharpoons H_3N{-}BH_3$$

NH_3 uses the lone pair of electrons on N (donor) to bond with B (acceptor) by forming a coordinate bond. So NH_3 is the base and BH_3 the acid.

The Bronsted–Lowry concept of acids and bases

A Bronsted–Lowry **acid** is a proton donor and a Bronsted–Lowry **base** is a proton acceptor.

In the reaction

$$HCl(aq) + H_2O(l) \rightleftharpoons H_3O^+(aq) + Cl^-(aq)$$

$$\text{acid}_1 \qquad \text{base}_2 \qquad \text{acid}_2 \qquad \text{base}_1$$

HCl is the acid since it donates a proton, and H_2O the base since it accepts a proton. In the reverse reaction, H_3O^+ is the acid and Cl^- the base. HCl and Cl^- are called a **conjugate acid–base pair**. Another conjugate acid–base pair is H_3O^+ and H_2O.

> An acid minus an H⁺ ion forms its conjugate base. Similarly an H⁺ ion added to a base gives its conjugate acid.

A **hydronium ion** (or **hydroxonium ion**), H_3O^+, is a proton (hydrogen ion, H^+) bonded to a water molecule. Though hydrogen ions exist in the form of hydronium ions in aqueous solutions, we often refer to them simply as hydrogen ions.

Water acts as an acid when it donates a proton to a base to form OH^-, and as a base when it accepts a proton from an acid to form H_3O^+. Water ionises to a very slight extent as shown below.

$$H_2O + H_2O \rightleftharpoons H_3O^+ + OH^-$$

$$\text{acid} \quad \text{base} \quad \text{acid} \quad \text{base}$$

> Remember that [ ] stands for concentration in mol dm⁻³.

Because the formation of one H_3O^{+1} ion is accompanied by the formation of one OH^- ion, in pure water and in neutral aqueous solutions, $[H_3O^+] = [OH^-] = 10^{-7}$ mol dm⁻³ at 25°C.

The ionic product of water, K_w

> Note that [H₂O] does not appear in this equilibrium expression as it is very large compared to [H₃O⁺] and [OH⁻] and is effectively constant.

The equilibrium constant for the ionisation of water is called the ionic product of water, K_w.

$$2H_2O(l) \rightleftharpoons H_3O^+(aq) + OH^-(aq)$$

$$K_w = [H_3O^+][OH^-]$$

Or simply, $K_w = [H^+][OH^-]$

Since $[H^+] = [OH^-] = 10^{-7}$ mol dm^{-3} at 25°C,
then $K_W = 10^{-14}$ mol^2 dm^{-6}.

Acids and the acid dissociation constant, K_a

A **strong acid** is one which ionises completely in solution.
Hydrochloric acid (HCl), hydrobromic acid (HBr) and
perchloric acid (HClO$_4$) are examples of strong acids.
A 1.0 mol dm^{-3} solution of HCl contains 1.0 mol dm^{-3} H$^+$
and 1.0 mol dm^{-3} Cl$^-$ ions (because of complete
ionisation).

A **weak acid** is one which ionises partially in solution.
Hydrocyanic acid (HCN) and ethanoic acid (CH$_3$COOH)
are examples of weak acids. A 1.0 mol dm^{-3} solution of
CH$_3$COOH has an H$^+$ concentration much smaller than
1.0 mol dm^{-3}. The concentration of H$^+$ ions in a weak acid
can be calculated using its dissociation constant.

For the ionisation of a weak acid, HA (A stands for the
anion of the acid),

$$HA(aq) + H_2O(l) \rightleftharpoons H_3O^+(aq) + A^-(aq)$$

the equilibrium constant is

$$K_a = \frac{[H_3O^+][A^-]}{[HA]}$$

K_a is the **acid dissociation constant** and it is a constant for a
given acid at constant temperature. The higher the value of
K_a, the stronger the acid is.

Another measure of acid strength is pK_a. **pK_a** is the negative
logarithm to the base 10 of K_a.

$$pK_a = -\log_{10} K_a$$

The lower the pK_a value of an acid, the stronger
the acid is.

Bases and the base dissociation constant, K_b

A **strong base** is one which ionises completely in aqueous solution. Sodium hydroxide (NaOH) and potassium hydroxide (KOH) are examples of strong bases. A 1.0 mol dm^{-3} solution of NaOH contains 1 mol dm^{-3} of both Na$^+$ and OH$^-$ ions.

A **weak base** is one which ionises partially in solution. Ammonium hydroxide (an aqueous solution of NH$_3$) and methylamine (CH$_3$NH$_2$) are examples of weak bases.

In the dissociation of a base B,

$$B(aq) + H_2O(l) \rightleftharpoons BH^+(aq) + OH^-(aq)$$

B accepts a proton from water. The base dissociation constant K_b is

$$K_b = \frac{[BH^+]\,[OH^-]}{[B]}$$

The higher the K_b value, the stronger the base is.

pK_b is the negative logarithm to the base 10 of K_b.

$$pK_b = -\log_{10} K_b$$

The lower the **pK_b** value of a base, the stronger the base is.

The relationship between acid and base dissociation constants

The product of K_a of an acid and K_b of its conjugate base is 10^{-14} at 25°C. (See Example 7.6)

$$K_a \times K_b = 10^{-14} \text{ mol}^2 \text{ dm}^{-6}$$

The sum of the pK_a of an acid and the pK_b of its conjugate base is also equal to 14.

$$pK_a + pK_b = 14$$

pH

This is a common measure used in acid–base chemistry. The **pH** is the negative logarithm to the base 10 of the hydrogen ion concentration of solution,

$$pH = -\log_{10}[H^+]$$

As discussed earlier, in pure water and neutral aqueous solutions $[H^+] = 10^{-7}$ mol dm^{-3} and so pH = 7.

Another useful relationship is shown by the ionic product of water.

$$[H^+]\,[OH^-] = 10^{-7} \times 10^{-7} = 10^{-14} \text{ mol}^2 \text{ dm}^{-6}$$

$$\text{So,} \quad pH + pOH = 14$$

The pH of all acidic solutions is lower than 7 and that of base (alkaline) solutions is higher than 7.

A **buffer solution** is one which resists pH change on the addition of small quantities of an acid or a base. A weak acid and its conjugate base (for example, CH_3COOH and CH_3COO^-) or a weak base and its conjugate acid (for example, NH_3 and NH_4^+) form buffer solutions. Example 7.10 shows how to calculate the pH of a buffer solution and Example 7.11 how to prepare a buffer solution of given pH.

Relationship between pH and pK_a

When a weak acid is in equilibrium with its conjugate base in solution, its pH and pK_a are related by the equation,

$$pH = pK_a + \log \frac{[\text{base}]}{[\text{acid}]}$$

7.2 Calculations based on the ionic product of water

Example 7.1 Calculating [H⁺] and [OH⁻] in acidic and alkaline solutions

Calculate $[H^+]$ and $[OH^-]$ in the following solutions.

(a)   0.01 mol dm^{-3} HCl

(b)   0.05 mol dm^{-3} NaOH

Solution

(a) HCl dissociates completely in solution.

$$HCl(aq) \rightarrow H^+(aq) + Cl^-(aq)$$

So in 0.01 mol dm^{-3} HCl,

$$[H^+] = 0.01 \text{ mol dm}^{-3}$$

$$[H^+][OH^-] = 10^{-14} \text{ mol}^2 \text{ dm}^{-6}$$

$$[OH^-] = \frac{10^{-14}}{10^{-2}} = 10^{-12} \text{ mol dm}^{-3}$$

(b) NaOH dissociates completely in solution. So in 0.05 mol dm^{-3} NaOH solution,

$$[OH^-] = 0.05 \text{ mol dm}^{-3}$$

$$[H^+][OH^-] = 10^{-14} \text{ mol}^2 \text{ dm}^{-6}$$

$$[H^+] = \frac{10^{-14}}{5 \times 10^{-2}} = 2.0 \times 10^{-13} \text{ mol dm}^{-3}$$

7.3 Calculation involving the pH of solutions of strong acids and bases

Example 7.2 Calculating pH from [H⁺]

You can work out the value of this on your calculator by entering `2 × 10⁻⁵` and then pressing the `log` button – but don't forget to reverse the sign of the displayed answer

Calculate the pH of a solution which contains 2.0×10^{-5} mol dm^{-3} of HNO$_3$.

Solution

HNO$_3$ dissociates completely to H$^+$ and NO$_3^-$ ions. So,

$$[H^+] = 2.0 \times 10^{-5} \text{ mol dm}^{-3}$$
$$pH = -\log_{10}[H^+]$$
$$= -\log 2.0 \times 10^{-5}$$
$$= 4.70$$

Example 7.3

Calculate the pH of each of the following solutions.

(a) 0.001 mol dm^{-3} HNO$_3$ (b) 0.02 mol dm^{-3} H$_2$SO$_4$

(c) 0.01 mol dm^{-3} KOH   (d) 0.05 mol dm^{-3} Ba(OH)$_2$

Solution

(a)
$$[H^+] = 10^{-3} \text{ mol dm}^{-3}$$
$$pH = -\log 10^{-3}$$
$$= 3$$

All the substances in this question ionise completely in solution.

(b) One H$_2$SO$_4$ molecule ionises to give two H$^+$ ions. So in 2.0×10^{-2} mol dm^{-3} H$_2$SO$_4$,

$$[H^+] = 4.0 \times 10^{-2} \text{ mol dm}^{-3}$$
$$pH = -\log 4.0 \times 10^{-2}$$
$$= 1.4$$

(c)
$$[OH^-] = 10^{-2} \text{ mol dm}^{-3}$$
$$[H^+] = \frac{10^{-14}}{10^{-2}} = 10^{-12} \text{ mol dm}^{-3}$$
$$pH = -\log 10^{-12}$$
$$= 12$$

(d)
$$[OH^-] = 2 \times 0.05 = 0.1 \text{ mol dm}^{-3}$$
$$[H^+] = \frac{10^{-14}}{10^{-1}} = 10^{-13} \text{ mol dm}^{-3}$$
$$pH = -\log 10^{-13}$$
$$= 13$$

Example 7.4 Calculating [H$^+$] from pH

The pH of a solution is 8.4. Calculate (a) the pOH (b) the [H$^+$] and (c) the [OH$^-$] of the solution.

Solution

(a)
$$pH = 8.4$$

$$pH + pOH = pK_w = 14$$

$$pOH = 14 - 8.4 = 5.6$$

(b)
$$pH = 8.4$$

$$[H^+] = 10^{-8.4}$$

$$= 10^{0.6} \times 10^{-9}$$

$$= 4.0 \times 10^{-9} \text{ mol dm}^{-3}$$

> You can work out the value of this on your calculator by entering '– 8.4' and the pressing the '10ˣ' button.

(c)
$$[OH^-] = \frac{10^{-14}}{[H^+]}$$

$$= \frac{10^{-14}}{4.0 \times 10^{-9}}$$

$$= 2.5 \times 10^{-6} \text{ mol dm}^{-3}$$

7.4 Calculations based on acid and base dissociation constants

Example 7.5 Comparing acid strengths

Arrange the acids whose acid dissociation constants are given below in order of increasing acid strength.

Acid	K_a(mol dm^{-3})
Hydrogen sulphide, H_2S	8.9×10^{-8}
Hypochlorous acid, $HOCl$	3.7×10^{-8}
Nitric acid, HNO_3	40.0
Ammonium ion, NH_4^+	5.6×10^{-10}

Solution

As K_a increases, the acid strength increases.

$$NH_4^+ < HOCl < H_2S < HNO_3$$
$$\text{acid strength increases} \longrightarrow$$

Example 7.6 Calculating the base dissociation constant, K_b, using the K_a value of the conjugate acid

The acid dissociation constant K_a of NH_4^+ is 5.6×10^{-10} mol dm⁻³. Calculate the base dissociation constant, K_b, of NH_3.

Solution

NH_3 is the conjugate base of the acid NH_4^+.

$$K_a \, (NH_4^+) \times K_b \, (NH_3) = 10^{-14} \, mol^2 \, dm^{-6}$$

$$K_b \, (NH_3) = \frac{10^{-14} \, mol^2 \, dm^{-6}}{5.6 \times 10^{-10} \, mol \, dm^{-3}}$$

$$= 1.8 \times 10^{-5} \, mol \, dm^{-3}$$

Example 7.7 Calculating pK_a from the acid dissociation constant, K_a

Given the K_a values below, calculate the pK_a values of benzoic acid and phenol. Which of the two is a stronger acid?

K_a(mol dm⁻³): benzoic acid, $C_6H_5COOH = 6.5 \times 10^{-5}$; phenol, $C_6H_5OH = 1.3 \times 10^{-10}$

Solution

$$pK_a(C_6H_5COOH) = -\log K_a = -\log 6.5 \times 10^{-5}$$

$$= 4.2$$

$$pK_a(C_6H_5OH) = -\log 1.3 \times 10^{-10}$$

$$= 9.9$$

Benzoic acid is a stronger acid than phenol.

7.5 Calculations involving the pH of solutions of weak acids

Example 7.8 Writing an expression for K_a and calculating the pH of a solution of a weak acid

A solution contains 0.1 mol dm^{-3} of ethanoic acid (CH_3COOH).

(a) Write an expression for the acid constant K_a.

(b) Calculate the pH of the solution. [$K_a(CH_3COOH)$ = 1.8×10^{-5} mol dm^{-3}]

Solution

(a)

$$CH_3COOH(aq) \rightleftharpoons CH_3COO^-(aq) + H^+(aq)$$

$$K_a = \frac{[CH_3COO^-]\,[H^+]}{[CH_3COOH]}$$

(b) Ethanoic acid is a weak acid as it dissociates only partially in solution. In a 0.1 mol dm^{-3} solution, if x mol of CH_3COOH dissociates, x mol each of CH_3COO^- and H^+ ions are formed.

$$CH_3COOH(aq) \rightleftharpoons CH_3COO^-(aq) + H^+(aq)$$

Equilibrium
concentration $(0.1 - x)$ x x (mol dm^{-3})

$$K_a = \frac{[CH_3COO^-]\,[H^+]}{[CH_3COOH]}$$

$$1.8 \times 10^{-5} = \frac{x^2}{0.1 - x}$$

Since x is small compared to 0.1, x can be neglected in the denominator and $(0.1 - x)$ is approximated to 0.1.

$$1.8 \times 10^{-5} = \frac{x^2}{0.1}$$

$$x^2 = 1.8 \times 10^{-6}$$

$$x = 1.34 \times 10^{-3}$$

Therefore, $[H^+] = 1.34 \times 10^{-3}$ mol dm^{-3}

$$pH = -\log 1.34 \times 10^{-3}$$

$$= 2.87$$

Example 7.9 Calculating K_a and pK_a values from pH

The pH of a 0.05 mol dm^{-3} solution of hydrofluoric acid is 2.3. Calculate the $[H^+]$ of the solution and the K_a and pK_a values of hydrofluoric acid.

Solution

Step 1: First, calculate the $[H^+]$ from the pH.

$$pH = 2.3$$

$$[H^+] = 10^{-2.3}$$

$$= 10^{0.7} \times 10^{-3}$$

$$= 5.0 \times 10^{-3} \text{ mol dm}^{-3} (0.005 \text{ mol dm}^{-3})$$

Step 2: Calculate the equilibrium concentrations of HF, H$^+$ and F$^-$ in the solution.

HF dissociates in solution to produce equal number of moles of H$^+$ and F$^-$.

$$HF(aq) \rightleftharpoons H^+(aq) + F^-(aq)$$

$$[F^-] = [H]^+ = 0.005 \text{ mol dm}^{-3}$$

Therefore, $[HF] = (0.05 - 0.005) = 0.045$ mol dm^{-3}

Step 3: Write an expression for K_a, substitute the equilibrium concentrations and calculate the values of K_a and then pK_a.

$$HF(aq) \rightleftharpoons H^+(aq) + F^-(aq)$$

Equilibrium
concentration: 0.045   0.005   0.005 (mol dm^{-3})

$$K_a = \frac{[H^+]\,[F^-]}{[HF]}$$

$$= \frac{0.005 \times 0.005}{0.045}$$

$$= 5.56 \times 10^{-4} \text{ mol dm}^{-3}$$

$$pK_a = -\log 5.56 \times 10^{-4}$$

$$= 3.3$$

7.6 Calculations involving buffer solutions

Example 7.10 Calculating the pH of a buffer solution

A buffer solution is made by dissolving 2.2 g of sodium ethanoate (CH_3COONa) in 500 cm^3 of ethanoic acid of concentration 0.02 mol dm^{-3}. Calculate the pH of the solution. (pK_a (CH_3COOH) = 4.75, M_R: CH_3COONa = 82)

Solution

Concentration of sodium ethanoate

$$= 2.20 \times \frac{1000}{500}$$

$$= 4.40 \text{ g dm}^{-3}$$

$$\text{Molar concentration} = \frac{4.40}{82}$$

$$= 0.054 \text{ mol dm}^{-3}$$

Sodium ethanoate is a soluble salt and dissociates completely in solution. Therefore,

$$[CH_3COONa] = [CH_3COO^-] = 0.054 \text{ mol dm}^{-3}$$

$$[CH_3COOH] = 0.02 \text{ mol dm}^{-3} \text{(given)}$$

$$pH = pK_a + \log \frac{[\text{base}]}{[\text{acid}]}$$

$$pH = pK_a + \log \frac{[CH_3COO^-]}{[CH_3COOH]}$$

$$= 4.75 + \log \frac{0.054}{0.02}$$

$$= 5.2$$

Example 7.11 Calculating quantities of substances needed to make a buffer solution of a given pH

Calculate the ratio of the number of moles of sodium methanoate (HCOONa) to that of methanoic acid (HCOOH) to prepare 1 dm³ of a buffer solution of pH 4.1. (K_a(HCOOH) = 1.6×10^{-4} mol dm^{-3}, pK_a(HCOOH) = 3.8)

Solution

$$pH = pK_a + \log \frac{[base]}{[acid]}$$

$$pH = pK_a + \log \frac{[HCOO^-]}{[HCOOH]}$$

$$4.1 = 3.8 + \log \frac{[HCOO^-]}{[HCOOH]}$$

$$\log \frac{[HCOO^-]}{[HCOOH]} = 0.3$$

$$\frac{[HCOO^-]}{[HCOOH]} = 10^{0.3} = 2.0$$

Therefore, the ratio of number of moles HCOO$^-$: HCOOH = 2:1. The number of moles of HCOO$^-$ should be double the number of moles of HCOOH to prepare a buffer solution of pH 4.1.

7.7 Practice questions

1. A dilute acid contains 0.05 mol dm^{-3} HCl. What is its H$^+$ ion concentration? What is the OH$^-$ ion concentration in the solution?

2. An alkaline solution contains 0.02 mol dm^{-3} Ba(OH)$_2$. Calculate the OH$^-$ and H$^+$ ion concentrations in the solution.

3. Calculate the [H$^+$] and [OH$^-$] in each of the following aqueous solutions. (Assume that all the acids and bases given below ionise completely.)

   (a) 2.5 mol dm^{-3} HNO$_3$

   (b) 0.2 mol dm^{-3} H$_2$SO$_4$

   (c) 1.0 × 10^{-3} mol dm^{-3} HCl

   (d) 1.0 × 10^{-3} mol dm^{-3} NaOH

   (e) 0.2 mol dm^{-3} KOH

   (f) 0.004 mol dm^{-3} Ca(OH)$_2$

4. Calculate the pH of each of the solutions in Question 3.

5. The pH of an aqueous solution of a weak acid is 5.3. Calculate its

   (a) [H$^+$]

   (b) [OH$^-$]

   (c) pOH

6. 100 cm^3 of 0.02 mol dm^{-3} HCl and the same volume of 0.04 mol dm^{-3} HNO$_3$ are mixed. Calculate the

   (a) [H$^+$] in the mixture

   (b) pH of the mixture.

7. The pOH of a NaOH solution is 1.5. Calculate its [OH$^-$] and [H$^+$].

8. Use the K_a values of the acids given below to arrange them in order of decreasing acid strength.

	K_a (mol dm^{-3})
Benzoic acid, C$_6$H$_5$COOH	6.5 × 10^{-5}
Ethanoic acid, CH$_3$COOH	1.8 × 10^{-5}

Hydrocyanic acid, HCN $\qquad$ 4.9×10^{-10}
Lactic acid, $CH_3CH(OH)COOH$ $\qquad$ 8.4×10^{-4}
Phenol, C_6H_5OH $\qquad$ 1.3×10^{-10}

9. Arrange the following acids in order of increasing acid strength.

	K_a (mol dm^{-3})
Chloroethanoic acid, $CH_2ClCOOH$	1.4×10^{-3}
Dichloroethanoic acid, $CHCl_2COOH$	5.0×10^{-2}
Trichloroethanoic acid CCl_3COOH	3.0×10^{-1}
Ethanoic acid, CH_3COOH	1.8×10^{-5}

10. Use the base dissociation constants of ammonia (NH_3), methylamine (CH_3NH_2) and aniline ($C_6H_5NH_2$) given below to answer the questions.

	K_b (mol dm^{-3})
NH_3	1.8×10^{-5}
CH_3NH_2	3.6×10^{-4}
$C_6H_5NH_2$	4.3×10^{-10}

(a) Which of the three is the strongest base?

(b) Which is the weakest base?

(c) Write an equation for the dissociation of methylamine in water.

(d) Write an expression for the base dissociation constant of methylamine.

(e) Write the formulae of the conjugate acids of these bases and arrange them in order of increasing acid strength.

11. The pH of a 0.1 mol dm^{-3} solution of a weak acid is 2.9. Use this information to calculate its K_a and pK_a values.

12. Calculate the pH of 0.1 mol dm^{-3} NH_3(aq). (K_b of NH_3 = 1.8×10^{-5} mol dm^{-3})

13. Calculate the pH of a buffer solution prepared by dissolving 2.1 g of sodium ethanoate (CH_3COONa) and

3.0 g of pure ethanoic acid (CH_3COOH) in water to form a 200 cm^3 solution. (K_a of CH_3COOH = 1.8×10^{-5} mol dm^{-3})

14. Calculate the pH of a solution which contains 0.2 mol dm^{-3} NH_4Cl and 0.3 mol dm^{-3} NH_3. (K_a of NH_4^+ = 5.6×10^{-10} mol dm^{-3})

8 **Formula and structure determination**

One-minute overview

A number of steps are involved in the identification of an unknown organic compound. The determination of the percentage composition and molecular mass of the compound leads to its molecular formula. The functional group(s) can be identified by chemical tests, as well as by noting the position of the peaks in the infrared spectrum of the compound. The ^1H nuclear magnetic resonance spectrum of the compound helps to identify the number of hydrogen atoms and the environment in which these atoms are located in the molecule. However, any single technique is insufficient. A brief description of infrared, mass and nuclear magnetic resonance spectroscopy is given in this chapter.

The worked examples include:
- calculation of percentage composition of a compound
- determination of empirical and molecular formulae
- use of mass spectroscopy, infrared and NMR data to identify compounds.

8.1 Important terms

The **percentage composition** of a compound is the mass of each element present expressed as the percentage of the total mass of the compound. For example, one mole of methane (CH_4) molecules contains one mole of carbon atoms and four moles of hydrogen atoms. In mass, 16.0 g of methane contains 12.0 g of carbon and 4.0 g of hydrogen. The percentage by mass of carbon is $\left(\frac{12.0}{16.0}\right) \times 100 = 75.0\%$ and that of hydrogen is $\left(\frac{4.0}{16.0}\right) \times 100 = 25\%$.

The **molecular formula** of a compound shows the numbers of atoms of each element in one molecule of the compound. The **empirical formula** of a compound gives the simplest ratio of atoms of each element in the compound. For example, the molecular formula of butane is C_4H_{10}. There are four carbon atoms and ten hydrogen atoms in one molecule of butane. The ratio of the numbers of carbon and hydrogen atoms = 4:10 = 2:5. Therefore the empirical formula of butane is C_2H_5. A molecular formula is either the same as the empirical formula or a simple multiple of it.

Molecular formula = (empirical formula)$_n$, where

$$n = \frac{\text{molecular mass}}{\text{empirical formula mass}}$$

8.2 Structure determination: mass, IR and NMR spectroscopy

The **structure** of an organic compound can be determined using information from chemical analysis and from spectroscopic techniques such as mass spectroscopy, infrared (IR) spectroscopy and nuclear magnetic resonance (NMR) spectroscopy.

Mass spectroscopy

A **mass spectrometer** is used to find the atomic and isotopic masses of elements and molecular masses of compounds. In a mass spectrometer, molecules of an organic compound are bombarded with high energy electrons and electrons are knocked out of molecules to produce positively charged ions. These ions are called *molecular ions* (**or parent ion, M**⁺). Molecular ions may break down to give positively charged fragments. These positively charged ions are separated by their *mass-to-charge ratio* (*m/z*), detected and recorded on a chart to form a *mass spectrum*. The position of a peak in the spectrum represents the mass of a positive ion, and the height of the peak its relative abundance. The masses of fragments give clues about various groups present in the compound. For example, a molecular ion (M^+) has the same mass as the parent molecule; (M–15) indicates a CH_3 group; (M–18) the loss of an H_2O molecule, (M–31), an OCH_3 group; (M–45), a COOH group and so on. (see Table 8.1)

Mass lost from molecular ion	Possible group	Name of group
15	CH_3	methyl group
17	OH	alcohol
18	H_2O	alcohol
29	CHO	aldehyde
	C_2H_5	ethyl group
31	CH_3O	methyl ester
43	C_3H_7	propyl group
	CH_3CO	acetyl group
45	COOH	carboxylic acid
	C_2H_5O	ethyl ester
57	C_4H_9	butyl group

Table 8.1
Mass losses from molecular ions
◀

Figure 8.1 shows the mass spectrum of n–hexane ($CH_3CH_2CH_2CH_2CH_2CH_3$). The peak at m/z 86 is that of the molecular ion (M^+) and the one at m/z 71 is for the ion (M–15)$^+$ formed by the loss of a methyl group. The peaks at m/z values of 57, 43, 29 and 15 correspond to the fragments $(C_4H_9)^+$, $(C_3H_7)^+$, $(C_2H_5)^+$ and $(CH_3)^+$ respectively.

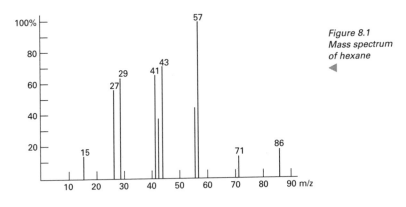

Figure 8.1
Mass spectrum of hexane
◀

Infrared spectroscopy

This is an important tool in the identification of functional groups such as –OH, –NH$_2$, –COOH, C=O, C=C in organic compounds. An infrared spectrometer is used to record the infrared spectrum of a compound. This is a plot of percentage transmission against wave number (usually between 4000 and 400 cm^{-1}). A typical IR spectrum is shown in Figure 8.2.

Figure 8.2 IR spectrum of butanone, CH$_3$CH$_2$COCH$_3$ ▶

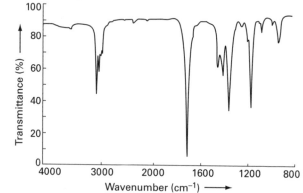

An IR spectrum can be quite complex with many 'downward' peaks. Each downward peak represents IR absorption of a

Table 8.2 Infrared absorption due to stretching vibrations ▶

Bond	Type of compound	Wave number (cm^{-1})
C–H	Alkanes, alkenes, arenes	2850 to 3100
O–H	Hydrogen-bonded alcohols and phenols	3200 to 3600
O–H	Hydrogen-bonded carboxylic acids	2500 to 3100
N–H	Primary amines	3300 to 3500
C=C	Alkenes	1610 to 1680
C≡C	Alkynes	2100 to 2150
C=O	Aldehydes, ketones, carboxylic acids and esters	1680 to 1750
C–O	Alcohols, ethers and esters	1000 to 1300
C–Cl	Halogenoalkanes	700 to 800

specific wave number. Most functional groups give rise to absorption peaks between wave numbers 3500 and 1600 cm^{-1}. The region between 1600 and 400 cm^{-1} is called the **fingerprint region**, which is unique for each compound. Table 8.2 gives some characteristic values for infrared absorption due to bond stretching vibrations in organic molecules.

NMR spectroscopy

This is one of the most valuable tools in structure determination. It provides information on the number of hydrogen atoms in a molecule as well as the chemical environment of these atoms. The NMR spectrum is calibrated using tetramethylsilane (TMS), $(CH_3)_4Si$, as the reference. The signal produced by the 12 equivalent H atoms of TMS is assigned an arbitrary chemical shift value of 0.0 ppm. The **chemical shift** (δ) is the position of a signal in the NMR spectrum and is expressed as parts per million (ppm) downfield from the TMS peak.

The chemical shifts of protons from different chemical environment are different (see Table 8.3).

Table 8.3
Proton chemical shift values

Environment of proton	Structure	Chemical shift (ppm)
Saturated primary	RCH_3	0.7 to 1.3
Saturated secondary	RCH_2R	1.2 to 1.4
Saturated tertiary	R_3CH	1.4 to 1.7
Alkyl ethanoate	CH_3COOR	2.0 to 2.1
Aryl ethanoate	CH_3COOAr	4.0 to 4.2
Alkyl chloride	RCH_2Cl	3.0 to 4.0
Methyl ether	$ROCH_3$	3.8 to 3.9
Methyl ketone	$RCOCH_3$	2.1 to 2.4
Aryl methyl	$ArCH_3$	2.2 to 2.5
Alcohol	ROH	0.5 to 6.0
Phenol	$ArOH$	7.0 to 7.2
Aromatic	ArH	6.5 to 8.0
Aldehyde	$RCHO$	9.7 to 10.0
Carboxylic acid	$RCOOH$	11.0 to 12.0

A low resolution NMR spectrum of ethanol, CH_3CH_2OH, shows three peaks – one at $\delta = 1.2$ ppm (CH_3 protons), one at $\delta = 3.6$ (CH_2 protons) and one at $\delta = 5.2$ (OH proton) – see Figure 8.3a. Note that the number of protons responsible for a peak is proportional to the area under the peak (and roughly proportional to the height of the peak).

Figure 8.3a Low resolution NMR spectrum of ethanol ▶

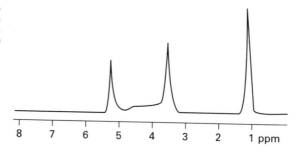

A high resolution spectrum of ethanol shows that the peaks at $\delta = 1.2$ ppm and $\delta = 3.6$ are not singlets but multiplets (Figure 8.3b).

Figure 8.3b High resolution NMR spectrum of ethanol ▶

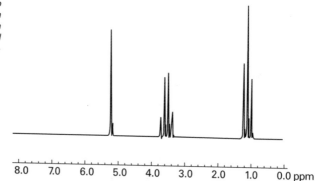

The areas under the set of peaks making up a multiplet add up to the area given by low resolution NMR analysis – computing this area is done in a mathematical procedure called integration (see Practice Question 12).

In a molecule, equivalent protons have the same chemical shift, whereas non-equivalent protons have different chemical shifts. For example, the two methyl groups (i.e., the six protons) of propanone, CH_3COCH_3, are equivalent and propanone gives a single signal at $\delta = 2.1$ ppm. The methyl groups in methyl ethanoate are non-equivalent and they produce two signals – a three-proton singlet at $\delta = 2.0$ ppm and another three-proton singlet at $\delta = 3.6$ ppm.

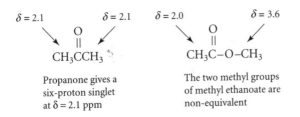

Propanone gives a six-proton singlet at $\delta = 2.1$ ppm	The two methyl groups of methyl ethanoate are non-equivalent

The signal of a proton splits into multiple signals called **multiplets** under the influence of neighbouring protons. Protons that have n neighbouring protons give rise to $(n + 1)$ peaks. For example butanone, $CH_3CH_2COCH_3$, produces three sets of peaks (Figure 8.4).

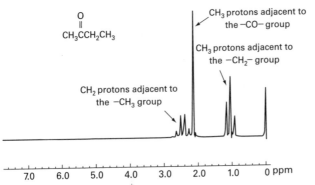

Figure 8.4
NMR spectrum of butanone

The peaks centred at $\delta = 2.44$ corresponding to $-CH_2-$ protons are a quartet (four peaks) because its signal is split by three CH_3 protons. The peaks centred at $\delta = 1.06$ corresponding to $CH_3 -$ protons are a triplet because of the coupling with the two CH_2 protons. The peak at $\delta = 2.15$ corresponding to $CH_3 -$ protons near the CO group is a singlet because there is no H atom attached to the neighbouring C atom and the signal is not split.

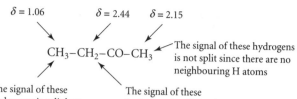

$$\delta = 1.06 \qquad \delta = 2.44 \qquad \delta = 2.15$$

$$CH_3-CH_2-CO-CH_3$$

The signal of these hydrogens is not split since there are no neighbouring H atoms

The signal of these hydrogens is split by two CH_2 protons ($n = 2$) to give a triplet ($n + 1$ peaks)

The signal of these hydrogens is split by three CH_3 protons ($n = 3$) to give a quartet ($n + 1$ peaks)

8.3 Calculations based on percentage composition and empirical formula

Example 8.1 Calculating percentage composition

Determine the percentage composition of ethanoic acid, CH_3COOH. (A_R: H = 1.0, C = 12.0, O = 16.0)

Solution

1 mole of CH_3COOH contains 2 moles of C atoms, 4 moles of H atoms and 2 moles of O atoms.

The molar mass of CH_3COOH is 60.0 g mol^{-1}.

60.0 g of ethanoic acid contains 24.0 g of C, 4.0 g of H and 32.0 g of O.

$$\% \text{ of C} = \frac{24.0 \times 100}{60.0} = 40.0\%$$

$$\% \text{ of H} = \frac{4.0 \times 100}{60.0} = 6.7\%$$

$$\% \text{ of O} = \frac{32.0 \times 100}{60.0} = 53.3\%$$

Example 8.2 Calculating percentage composition from combustion data

2.0 g of an organic acid containing C, H and O on combustion gave 1.96 g of CO_2 and 0.4 g of H_2O. Calculate the percentage composition of the acid.

Solution

During combustion, all the carbon present in the organic compound is converted to carbon dioxide and all the hydrogen to water.

Molar mass of CO_2 = 44.0 g mol^{-1}
44.0 g of CO_2 contains 12.0 g of C
1.96 g of CO_2 contains

$$\frac{12.0 \times 1.96}{44.0} = 0.53 \text{ g of C}$$

2.0 g of the organic compound contains 0.53 g of C.

$$\% \text{ of C} = \frac{0.53 \times 100}{2.0} = 26.50\%$$

Molar mass of H_2O = 18 g mol^{-1}
18.0 g of H_2O contains 2.0 g of H
0.4 g of H_2O contains

$$= \frac{2.0 \times 0.4}{18.0} = 0.044 \text{ g of H}$$

2.0 g of the organic compound contains 0.044 g of H

$$\% \text{ of H} = \frac{0.044 \times 100}{2.0} = 2.22\%$$

The remaining proportion of the compound must represent the oxygen present.

$$\% \text{ of O} = 100 - (26.50 + 2.22)$$
$$= 71.28\%$$

Example 8.3 Determining the empirical formula of a compound

Aniline (phenylamine) is an organic base. Aniline is found to contain 77.42% by mass of C, 7.53% by mass of H and 15.05% by mass of N. Work out the empirical formula of aniline.

Solution

100 g of aniline contains 77.42 g of C, 7.53 g of H and 15.05 g of N. To find the simplest ratio of the number of atoms in the molecule, follow these steps.

Step 1: Divide the percentages of elements by the respective atomic masses to get the mole ratio.

Step 2: Divide the numbers obtained in Step 1 by the smallest of them.

Step 3: Round off the answers to get simple whole numbers.

$$C \quad : \quad H \quad : \quad N$$

$$\frac{77.42}{12.0} \quad : \quad \frac{7.53}{1.0} \quad : \quad \frac{15.05}{14.0}$$

$$6.45 \quad : \quad 7.53 \quad : \quad 1.075$$

$$\frac{6.45}{1.075} \quad : \quad \frac{7.53}{1.075} \quad : \quad \frac{1.075}{1.075}$$

$$6 \quad : \quad 7 \quad : \quad 1$$

The empirical formula of aniline is C_6H_7N.

8.4 Calculations based on empirical and molecular formulae

Example 8.4 Deducing empirical and molecular formulae

An organic compound contains 53.33% C, 11.11% H and the remainder is O. The relative molecular mass of the compound is 90. Deduce the empirical and molecular formulae of the compound.

Solution

$$\% \text{ of O} = 100 - (53.33 + 11.11) = 35.56$$

$$C \quad : \quad H \quad : \quad O$$

$$\frac{53.33}{12.0} \quad : \quad \frac{11.11}{1.0} \quad : \quad \frac{35.56}{16.0}$$

$$4.44 \quad : \quad 11.11 \quad : \quad 2.22$$

$$\frac{4.44}{2.22} \quad : \quad \frac{11.11}{2.22} \quad : \quad \frac{2.22}{2.22}$$

$$2 \quad : \quad 5 \quad : \quad 1$$

Empirical formula = C_2H_5O

Molecular formula = (empirical formula)n

$$n = \frac{\text{molecular mass}}{\text{empirical formula mass}}$$

$$n = \frac{90}{45} = 2$$

Molecular formula = $(C_2H_5O)_2$

$$= C_4H_{10}O_2$$

8.5 Problems on structure determination

Example 8.5 Identifying molecular mass and structure from mass spectrometry data

An unknown organic compound, on chemical analysis, was found to be an ester. The mass spectrum of the compound showed peaks at m/z values of 15, 31, 43 and 74. What is the probable relative molecular mass of the compound? Identify the fragments and hence deduce the structural formula of the ester.

Solution

The general formula of an ester is RCOOR′. In the formula, R′ is an alkyl group and R can be an alkyl group or an H atom. Assuming that the highest m/z value is for the molecular ion, the relative molecular mass of the compound is 74.

The relative mass of the –COO– group is 44. The remaining mass (74 – 44) is the mass of the alkyl group(s).

The m/z value of 15 indicates a methyl group (CH_3), and 31 indicates a methoxy group (OCH_3). The fragment with m/z value 43 must be CH_3CO [CH_3 (mass 15) attached to CO (mass 28)].

Therefore, the structural formula of the ester is CH_3COOCH_3.

Example 8.6 Determining structural formula using percentage composition and spectral data

An organic compound contains 66.7% C, 11.1% H and 22.2% O. Its mass spectrum shows the molecular ion at $m/z = 72$. The IR spectrum of the compound contains a strong peak at 1715 cm^{-1}. The NMR spectrum of the compound shows a three-proton triplet centred at $\delta = 1.06$ ppm, a three-proton singlet at $\delta = 2.15$ ppm and a two-proton quartet centred at $\delta = 2.44$ ppm. Find the structural formula of the compound.

Solution

First, deduce the empirical formula of the compound and then the molecular formula.

$$
\begin{array}{ccccc}
C & : & H & : & O \\[4pt]
\dfrac{66.7}{12.0} & : & \dfrac{11.1}{1.0} & : & \dfrac{22.2}{16.0} \\[10pt]
5.56 & : & 11.1 & : & 1.39 \\[6pt]
\dfrac{5.56}{1.39} & : & \dfrac{11.1}{1.39} & : & \dfrac{1.39}{1.39} \\[10pt]
4 & : & 8 & : & 1
\end{array}
$$

Empirical formula = C_4H_8O

Empirical formula mass = 72

Molecular mass = 72 (mass spectrum data)

Therefore, molecular formula = C_4H_8O

The presence of one O atom in the molecule and a strong peak at 1715 cm^{-1} in the IR spectrum suggests a carbonyl (C=O) group (mass 28).

A three-proton singlet in the NMR spectrum suggests a methyl group (CH_3) attached to a carbon atom carrying no hydrogen atom. A three proton triplet shows a CH_3 group adjacent to two equivalent H atoms, and a two-proton quartet suggests a CH_2 group adjacent to three equivalent H atoms. This suggests the presence of an ethyl group (CH_3CH_2) in the molecule.

The structural formula of the compound is likely to be $CH_3CH_2COCH_3$. If so, the compound is butanone, a ketone of molecular mass 72 with a CH_3CH_2 group and a CH_3 group attached to a CO group.

8.6 Practice questions

Use these relative atomic masses for the following calculations.

A_R: H = 1.0, C = 12.0, N = 14.0, O = 16.0, Cl = 35.5

1. Calculate the percentage composition by mass of propanoic acid, molecular formula $C_3H_6O_2$.

2. Calculate the percentage composition by mass of phenol, C_6H_5OH.

3. Write the empirical formulae of these compounds.

    (a) Ethanedioic acid, $C_2H_2O_4$

    (b) Butanoic acid, $C_4H_8O_2$

    (c) Glucose, $C_6H_{12}O_6$

    (d) Sucrose, $C_{12}H_{22}O_{11}$

4. The ratio of carbon atoms to hydrogen atoms in an unsaturated hydrocarbon is 1:2. The relative molecular

mass of the compound is 84. Deduce the empirical and molecular formulae of the compound.

5. The ratio of carbon to hydrogen atoms in a hydrocarbon is 1:1 and its relative molecular mass is 78. Deduce the empirical and molecular formulae of the compound.

6. A compound was found to contain 60.0% C, 13.3% H and 26.7% O. Deduce its empirical formula. If the relative molecular mass of the compound is 60, what is its molecular formula?

7. An organic compound contains 40.0% C, 6.7% H and 53.3% O. Deduce its empirical formula. If the relative molecular mass of the compound is 180, what is its molecular formula?

8. An organic base of relative molecular mass 45 contains 53.3% C, 15.6% H and 31.1% N. Work out the empirical and molecular formulae of the compound.

9. The percentage composition of a halogenohydrocarbon of relative molecular mass 64.5 is: C = 37.2%; H = 7.8%; Cl = 55.0%. Determine the empirical and molecular formulae of the compound.

10. (a) Determine the empirical formula of an organic compound having the percentage composition by mass: C = 72.0%; H = 12.0%; O = 16.0%. The mass spectrum of the compound shows peaks at m/z values of 100, 85, 71, 58 and 43.

   (b) If the highest m/z value is that of the molecular ion, what is the molecular formula of the compound?

   (c) The IR spectrum of the compound indicates the presence of a carbonyl group (C=O). Deduce a probable structural formula of the compound.

11. The mass spectrum of an organic compound shows peaks at m/z values of 72, 57, 44, 43, 41 and 29. Chemical analysis confirms the presence of an aldehyde group (–CHO) in the compound. Suggest fragment ions

responsible for the peaks in the mass spectrum and work out a possible structural formula of the compound.

12. The mass spectrum of an organic compound of molecular formula $C_3H_6O_2$ shows peaks at m/z values of 74, 59, 43, 42 and 28. The NMR spectrum of the compound shows two singlets – one at $\delta = 2.0$ ppm (integrates to indicate 3H) and the other at $\delta = 3.6$ ppm (integrates to indicate 3H). The IR spectrum shows absorption bands at 1746 and 1246 cm^{-1}. Identify

   (a) as many fragment ions as possible in the mass spectrum,

   (b) the groups responsible for the peaks in the NMR spectrum,

   (c) the groups responsible for the absorption bands in the IR spectrum.

   Hence deduce the structure of the compound.

13. An organic compound of molecular formula $C_4H_8O_2$ gives a strong absorption band at 1747 cm^{-1} and another at 1247 cm^{-1} in the IR spectrum. The NMR spectrum of the compound contains a three-proton singlet at $\delta = 2.0$ ppm, a three-proton triplet at $\delta = 1.2$ ppm and a two-proton quartet at $\delta = 4.2$ ppm. Deduce the structure of the compound.

14. The molecular formula of an organic compound is $C_5H_{10}O$. In the IR spectrum of the compound, there is a strong absorption band at 1720 cm^{-1}. Its mass spectrum shows peaks at m/z values of 86, 57 and 29. The NMR spectrum gives a quartet centred at $\delta = 2.44$ ppm which integrates to indicate four protons and a triplet centred at $\delta = 1.06$ ppm which integrates to indicate six protons. Identify the compound.

15. The NMR spectrum of a pleasant smelling organic liquid of molecular formula $C_8H_8O_2$ shows a singlet (integrates to indicate 3H) at $\delta = 3.8$ ppm and a multiplet (integrates to indicate 5H) at $\delta = 7.2$–8.1 ppm.

On hydrolysis the compound gives a solid carboxylic acid which is sparingly soluble in water but dissolves in an alkali. Deduce the structure of the compound.

16. The mass spectrum of an organic liquid of molecular formula C_8H_8O shows peaks at m/z values of 120, 105, 77 and 43. The IR spectrum of the compound has an absorption peak at 1685 cm^{-1}. The NMR spectrum of the compound shows a three-proton singlet at $\delta = 2.4$ ppm and a five proton multiplet at $\delta = 7.2$–8.0 ppm.

(a) Identify the fragment ions responsible for the peaks in the mass spectrum.

(b) Deduce the structure of the compound.

Glossary

Activated complex   An intermediate formed from the reactant molecules which can change into products or reverse to the reactants

Activation energy, E_a   The minimum amount of energy needed for the formation of an activated complex

Anion   A negatively charged ion

Anode   The positively charged electrode which takes electrons from the negative ions attracted to it – it is the electrode at which oxidation occurs

Atomic mass unit, u   One-twelfth the mass of one atom of ^{12}C

Atomic number, Z   The number of protons in the nucleus of an atom

Avogadro constant, L   The number of atoms present in 12 g of ^{12}C ($L = 6.022 \times 10^{23}$)

Avogadro's law   At constant temperature and pressure, the volume of a gas is directly proportional to the number of moles of the gas

Bond enthalpy   The energy required to break the bonds in 1 mole of gaseous diatomic molecules into gaseous atoms

Boyle's law   The volume of a fixed mass of gas is inversely proportional to its pressure at constant temperature

Bronsted–Lowry acid   A proton acceptor

Bronsted–Lowry base   A proton donor

Buffer solution   A solution which resists any pH change for the addition of small quantities of an acid or a base – a weak acid and its conjugate base, or a weak base and its conjugate acid usually form buffer solutions.

Cathode	The negative electrode which the positive ions take electrons from – it is the electrode at which reduction takes place
Cation	A positively charged ion
Charles' law	At constant pressure, the volume of a fixed mass of gas is directly proportional to its temperature in Kelvin
Chemical kinetics	The study of rates of reactions
Collision theory	States that reactant molecules must collide before a reaction occurs, and that the rate of a reaction is proportional to the number of effective collisions between the reactant molecules
Combustion	A chemical reaction in which a substance burns in air or oxygen. For example, organic substances combines with oxygen to form carbon dioxide and water
Conjugate oxidant–reductant pair	An oxidant and the corresponding reduced form
Coulomb, C	The quantity of electricity passed when a current of one ampere is passed for one second
Dalton's law of partial pressures	The total pressure of a mixture of gases is equal to the sum of partial pressures of the gases

Disproportionation	A redox reaction in which one element undergoes oxidation and reduction simultaneously
Electrochemical cell	A set of apparatus in which a redox reaction is carried out in order to produce an electric current
Electrode	A graphite or metal rod used to carry electricity to an electrolyte
Electrolysis	A process by which chemical reactions occur due to the passage of electricity through an electrolyte
Electrolyte	A substance that conducts electricity in the dissolved form or molten state and undergoes chemical decomposition
Electrolytic cell	A set of apparatus consisting of two electrodes placed in an electrolyte in which an electric current is used to bring about chemical change
Electron	A negatively charged subatomic particle of negligible mass
Electron affinity	The energy charge when an electron is added to each atom in 1 mole of gaseous atoms
Empirical formula	A formula which shows the relative numbers of atoms of each element in a compound
Endothermic reaction	A reaction that requires energy to be input
Enthalpy, H	A term used to express the quantity of energy transferred at constant pressure during a reaction

Enthalpy change, ΔH	The difference between the sum of the enthalpies of the products and the sum of the enthalpies of the reactants
Enthalpy diagram	A diagram showing the relative levels of the enthalpies of reactants and products
Enthalpy of atomisation, ΔH_{atm}	The amount of energy required for the formation of 1 mole of gaseous atoms from the element in its standard physical state
Enthalpy of combustion, $\Delta H_c^{\ominus}$	The standard enthalpy of combustion is the energy evolved when one mole of a substance is burned completely in air or oxygen at 298 K and 1 atmosphere
Enthalpy of formation $\Delta H_f^{\ominus}$	The standard enthalpy of formation is the enthalpy change when 1 mole of a substance is formed form its constituent elements at 298 K and 1 atmosphere
Enthalpy of neutralisation	The energy evolved when 1 mole of H^+ ions is neutralised completely by 1 mole of OH^- ions
Enthalpy of reaction $\Delta H_{rxn}^{\ominus}$	The enthalpy change when the number of moles of the reactants in a stoichiometric equation react completely to form the products at 298 K and 1 atmosphere
Equilibrium system	A system in which forward and reverse reactions take place at the same rate
Exothermic reaction	A reaction that evolves energy

Faraday's constant	The charge carried by 1 mole of electrons – its value is 96500 coulombs
Faraday's law of electrolysis	The number of moles of a product formed during electrolysis is stoichiometrically equivalent to the amount of electricity (number of moles of electrons) supplied
Half-life, $t_{\frac{1}{2}}$	The time taken for the concentration of a substance to fall to half its original value
Hess's law	Whether a reaction takes place in one step or through a series of steps, the total energy change remains the same
Heterogeneous catalyst	A catalyst which is in a different phase (state) from the reactants – for example V_2O_5 (s) in the preparation of SO_3 (g) from SO_2 (g) and O_2 (g)
Heterogeneous system	A system in which at least one substance is in a different phase (state) from that of the other substances
Homogeneous catalyst	A catalyst which is in the same phase (state) as the reactants – for example H_2SO_4 (aq) in the hydrolysis of methyl methanoate, $HCOOCH_3$ (aq)
Homogeneous system	A system in which all the reactants and products are in the same phase (state)
Ionic product of water, K_w	The equilibrium constant for the ionisation of water
Ionisation energy	The energy required to remove an electron from each atom of one mole of gaseous atom

Isotopes	Atoms of the same element with the same number of protons but different numbers of neutrons, hence different masses
Lattice energy, ΔH_{lat}	The energy evolved when 1 mole of a solid ionic compound is formed from its gaseous ions under standard conditions
Le Chatelier's principle	If a stress (such as changing concentration, pressure or temperature) is applied to a system at equilibrium, the system adjusts in a direction so as to cancel the effect of the change
Lewis acid	A species which can accept a pair of electrons
Lewis base	A species which can donate a pair of electrons
Mass number	The total number of protons and neutrons in an atom
Mass spectrometry	A technique for determining the atomic and isotopic masses of elements and molecular masses of compounds, and their relative abundance, by passing positive ions through a magnetic field
Molar mass	The mass of 1 mole of a substance – molar mass is expressed in $g\,mol^{-1}$
Molar volume	The volume occupied by 1 mole of a substance – molar volume is constant for all *gases* at constant temperature and pressure and is 24.0 dm^3 at 1 atm pressure and 293 K

Molarity	The number of moles of a solute present in 1 dm^3 of a solution
Molecular formula	A formula which gives the numbers of atoms of each element in one molecule of a compound
Molecularity	The number of chemical species reacting in an elementary step in a reaction mechanism
Mole fraction	The number of moles of a component in a mixture divided by the total number of moles in the mixture
Neutron	A neutral subatomic particle of mass 1u
Order of a reaction	The sum of the exponents of the concentration terms in the overall rate equation for a reaction
Oxidation	(1) A reaction in which a substance loses electrons (2) A reaction in which the oxidation number of an element is increased
Oxidation number	The number of charges an atom would have in a substance if electrons were transferred completely
Oxidising agent (oxidant)	(1) An electron acceptor (2) A substance that decreases in oxidation number during a redox reaction
Partial pressure	The pressure exerted by a gas in a container if it alone were present in the container at the same temperature

Percentage composition	The mass of an element expressed as the percentage of the total mass of the compound
Periodic table	An arrangement of elements in groups (vertical columns) and periods (horizontal rows) based on atomic number and electronic configuration in a way to show the relationship between the properties of the elements
Proton	A positively charged subatomic particle of mass $1u$
Rate-determining step	If a reaction takes place through a number of steps, the slowest step is the rate-determining step
Rate of reaction	The amount of a substance reacted (mol dm^{-3}), or of a product formed in unit time (usually 1 second). The unit for rate of reaction is $\text{mol dm}^{-3}\,\text{s}^{-1}$
Reaction mechanism	The sequence of proposed elementary steps which lead to the formation of products
Redox reaction	A reaction in which oxidation and reduction take place simultaneously
Reducing agent (reductant)	(1) An electron donor (2) A substance that increases in oxidation number during a redox reaction
Reduction	(1) A reaction in which a substance gains electrons (2) A reaction in which the oxidation number of an element is decreased

Relative atomic mass, A	The mass of an atom of an element in relation to the mass of an atom of ^{12}C
Relative molecular mass, M	The mass of molecule of a substance in relation to the mass of an atom of ^{12}C
Standard cell potential ($E_{cell}^{\ominus}$)	The sum of standard electrode potentials for the oxidation and reduction reaction that take place in a cell
Standard electrode potential $E^{\ominus}$	The standard electrode potential of an electrode is the potential difference between that electrode and a hydrogen electrode, both electrodes under standard conditions. Standard electrode potential of hydrogen is assigned the value 0.00 V
Stoichiometric equation	A balanced chemical equation
Thermochemical equation	A stoichiometric equation for a reaction which also states the associated enthalpy change for the reaction
Thermochemistry	The study of heat changes in chemical reactions

Physical constants

$1\ dm^3 = 1\ litre = 10^3\ cm^3$

$1\ m^3 = 10^3\ dm^3 = 10^6\ cm^3$

Avogadro constant $= 6.022 \times 10^{23}\ mol^{-1}$

1 Faraday $= 96500\ C\ mol^{-1}$

Gas constant $R = 8.314\ J\ K^{-1}\ mol^{-1}$
$= 0.0821\ dm^3\ atm\ K^{-1}\ mol^{-1}$

$1\ atm = 1.01 \times 10^5\ N\ m^{-2}$

$1\ N\ m^{-2} = 1\ Pa$

Molar volume of a gas at 273 K and 1 atm = 22.4 dm^3
Molar volume of a gas at 298 K and 1 atm = 24.0 dm^3

Specific heat capacity of water = 4.18 J g^{-1} K^{-1}
$\qquad\qquad\qquad\qquad\qquad$ = 4.18 kJ kg^{-1} K^{-1}

Ionic product of water, $K_w = 10^{-14}$ mol^2 dm^{-6} at 298 K

Some relevant equations

$$\frac{p_1 V_1}{T_1} = \frac{p_2 V_2}{T_2}$$

$$pV = nRT$$

$$t_{\frac{1}{2}} = \frac{0.693}{k}$$

$$k = A\ e^{\frac{-Ea}{RT}}$$

$\ln k = \ln A - \frac{Ea}{RT}$

$\Delta H^{\ominus}_{rxn} = \Sigma\ \Delta H^{\ominus}$ (products) $-\ \Sigma\ \Delta H^{\ominus}$ (reactants)

$\Delta H_{rxn} = \Sigma\ BE$ (products) $-\ \Sigma\ BE$ (reactants)

$pH = -\log_{10} [H^+]$

$pK = -\log_{10} K$

Answers to practice questions

Chapter 1

1. 24.306

2. 35.455

3. (a) 286.0 g mol^{-1}

   (b) 180.0 g mol^{-1}

   (c) 136.9 g mol^{-1}

   (d) 40.0 g mol^{-1}

   (e) 80.1 g mol^{-1}

   (f) 38.0 g mol^{-1}

   (g) 96.1 g mol^{-1}

   (h) 131.6 g mol^{-1}

   (i) 329.2 g mol^{-1}

4. (a) 2.5 mol

   (b) 0.20 mol

   (c) 0.058 mol

   (d) 0.02 mol

5. (a) 8.78 g

   (b) 24.97 g

   (c) 3.20 g

   (d) 2.20 × 10^3 g

6. (a) 9.02 × 10^{22} molecules

   (b) 1.20 × 10^{23} molecules

7. 22.44 g of CaO

8. 4.71 g of $Cu(NO_3)_2$

9. 1.2 dm^3 of CO_2

10. (a) 0.09 mol of ions (0.03 mol of Mg^{2+} and 0.06 mol of Cl^- ions)

    (b) 0.30 mol dm^{-3}

11. 17.1 cm^3 of concentrated HCl

12. 1.0 dm^3 of water

13. (a) 0.10 mol of MgO

    (b) 0.05 mol of HCl

    (c) 0.075 mol of MgO

14. 250 cm^3 of H_2SO_4

15. (a) 1.5×10^{-3} mol of HCl

    (b) 7.5×10^{-4} mol of $Ca(OH)_2$

    (c) 0.015 mol dm^{-3}

    (d) 1.11 g dm^{-3}

16. (a) 0.04 mol of HCl

    (b) 0.04 mol of $NaHCO_3$

    (c) 3.36 g of $NaHCO_3$

    (d) 42.0%

Chapter 2

1. (a) 5.0 dm^3

    (b) 50.0 dm^3

    (c) 11.2 dm^3

    (d) 1.8 dm^3

2. 838.6 cm^3

3. 1456.0 cm^3

4. 113.7 kN m^{-2}

5. 27398.4 N m^{-2}

6. 2.05 atm

7. 6.2 × 10^4 N m^{-2}

8. 0.5 mol

9. 72.0

10. 59.9

Chapter 3

1. (a) endothermic

    (b) endothermic

    (c) exothermic

    (d) endothermic

    (e) exothermic

2. (a) −726.4 kJ mol^{-1}

    (b) 113.5 kJ

3. ΔH_f (C$_2$H$_4$, g) = +52.2 kJ mol^{-1}

4. ΔH_f (C$_6$H$_6$, g) = +49.0 kJ mol^{-1}

5. ΔH_c (C$_3$H$_8$, g) = −2219.2 kJ mol^{-1}

6. +7.2 kJ mol^{-1}

7. ΔH_{rxn} = −121.6 kJ

8. C–C bond energy = +347.0 kJ mol^{-1}

9. ΔH_{rxn} = −20.0 kJ mol^{-1}

10. I–Cl bond energy = +151.5 kJ mol^{-1}

11. ΔH_{lat}(CaO, s) = −3454.5 kJ mol^{-1}

12. ΔH_f^0 (MgI$_2$, s) = −364 kJ mol^{-1}

Chapter 4

1.

(a) $K_c = \dfrac{[ClF]^2}{[Cl_2][F_2]}$ (no unit)

(b) $K_c = \dfrac{[NO_2]^2}{[NO]^2[O_2]}$ mol^{-1} dm^3

(c) $K_c = \dfrac{[CO][H_2]}{[H_2O]}$ mol dm^{-3}

(d) $K_c = \dfrac{[HCOOH][CH_3OH]}{[HCOOCH_3][H_2O]}$ (no unit)

2.

(a) $K_p = \dfrac{p_{NO}^2}{p_{N_2} \times p_{O_2}}$ (no unit)

(b) $K_p = \dfrac{p_{NO_2}^2}{p_{N_2O_4}}$ atm

(c) $K_p = \dfrac{p_{HI}^2}{p_{H_2} \times p_{I_2}}$ (no unit)

(d) $K_p = \dfrac{p_{SO_2}^2 \times p_{O_2}}{p_{SO_3}^2}$ atm

3. 1.04

4. 1.44

5. 156.25

6. 0.25

7. 23.7 mol dm^{-3}

8. 0.05 mol dm^{-3}

9. (a) $CH_3COOH(l) + C_2H_5OH(l) \rightleftharpoons CH_3COOC_2H_5(l)$
$+ H_2O(l)$

(b) $K_c = \dfrac{[CH_3COOC_2H_5]\,[H_2O]}{[CH_3COOH]\,[C_2H_5OH]}$

(c) 4.12

10. 0.23

11. 0.952 atm

Chapter 5

1. (a) Order with respect to A = 1 and with respect to B = 1. Overall order = 2.

   (b) Order with respect to AX = 2. Overall order = 2.

   (c) Order with respect to H_2 = 1 and with respect to NO = 2. Overall order = 3.

2. (a) 2.0×10^{-7} mol dm^{-3} s^{-1}

   (b) 1.0×10^{-7} mol dm^{-3} s^{-1}

   (c) 2.0×10^{-7} mol dm^{-3} s^{-1}.

3. Rate of reaction would be:

   (a) increased by four times,

   (b) increased by two times,

   (c) decreased by half,

   (d) increased by eight times.

4. Rate = $k\,[Y]^2$

5. (a) Rate = $k\,[CH_3CHO]^2$

   (b) $k = 2.0$ mol^{-1} dm^3 s^{-1}

   (c) 0.5 mol dm^{-3} s^{-1}

6. (a) Order with respect to XO_3^- = 1, with respect to X^- = 1 and with respect to H^+ = 2,

   (b) Rate = $k\,[XO_3^-]\,[X^-]\,[H^+]^2$,

(c) $k = 1.3 \times 10^{-2} \text{ mol}^{-3} \text{ dm}^9 \text{ s}^{-1}$,

(d) Initial rate of reaction $= 1.3 \times 10^{-6} \text{ mol dm}^{-3} \text{ s}^{-1}$.

7. $N_2O \rightarrow N_2 + O$ slow

     $N_2O + O \rightarrow N_2 + O_2$ fast

8. Rate $= k \, [NO]^2 \, [H_2]$

Chapter 6

1. (a) P_2O_3,   P $= +3$, O $= -2$

     P_2O_5,   P $= +5$, O $= -2$

     CO,   C $= +2$, O $= -2$

     CO_2,   C $= +4$, O $= -2$

     SiO_2,   Si $= +4$, O $= -2$

     BaO,   Ba $= +2$, O $= -2$

     Al_2O_3,   Al $= +3$, O $= -2$

     Cr_2O_3,   Cr $= +3$, O $= -2$

(b) $NaHCO_3$,   Na $= +1$, H $= +1$, C $= +4$,   O $= -2$

     $KMnO_4$,   K $= +1$, Mn $= +7$, O $= -2$

     $CaCl_2$,   Ca $= +2$, Cl $= -1$

     AgCl,   Ag $= +1$, Cl $= -1$

     $CuBr_2$,   Cu $= +2$, Br $= -1$

     $Ba(NO_3)_2$,   Ba $= +2$, N $= +5$, O $= -2$

     $MgSO_4$,   Mg $= +2$, S $= +6$, O $= -2$

     Na_2S,   Na $= +1$, S $= -2$

2. NO_3^-,   N $= +5$, O $= -2$

     SO_3^{2-},   S $= +4$, O $= -2$

     CrO_4^{2-},   Cr $= +6$, O $= -2$

HSO_4^-, $H = +1$, $S = +6$, $O = -2$

HPO_4^{2-}, $H = +1$, $P = +5$, $O = -2$

HCO_3^-, $H = +1$, $C = +4$, $O = -2$

OH^-, $O = -2$, $H = +1$

SiO_3^{2-}, $Si = +4$, $O = -2$

3. KNO_2, $K = +1$, $N = +3$, $O = -2$

$NaNO_3$, $Na = +1$, $N = +5$, $O = -2$

HNO_3, $H = +1$, $N = +5$, $O = -2$

N_2O_5, $N = +5$, $O = -2$

N_2O, $N = +1$, $O = -2$

NO, $N = +2$, $O = -2$

N_2O_4, $N = +4$, $O = -2$

NO_2, $N = +4$, $O = -2$

4. SO_2, $S = +4$, $O = -2$

SO_3, $S = +6$, $O = -2$

$NaHSO_3$, $Na = +1$, $H = +1$, $S = +4$, $O = -2$

H_2SO_4, $H = +1$, $S = +6$, $O = -2$

Na_2S, $Na = +1$, $S = -2$

$ZnSO_4$, $Zn = +2$, $S = +6$, $O = -2$

$KHSO_4$, $K = +1$, $H = +1$, $S = +6$, $O = -2$

H_2SO_3, $H = +1$, $S = +4$, $O = -2$

5. (a) $Ba \rightarrow Ba^{2+} + 2e^-$

(b) $2Br^- \rightarrow Br_2 + 2e^-$

(c) $F_2 + 2e^- \rightarrow 2F^-$

(d) $Cu^+ \rightarrow Cu^{2+} + e^-$

(e) $S + 2e^- \rightarrow S^{2-}$

6. (a) $\qquad$ $Na \rightarrow Na^+ + e^-$ oxidation

$\qquad Br_2 + 2e^- \rightarrow 2Br^-$ reduction

$\qquad 2Na + Br_2 \rightarrow 2Na^+ + 2Br^-$ overall

(b) $\qquad$ $2I^- \rightarrow I_2 + 2e^-$ oxidation

$\qquad Cl_2 + 2e^- \rightarrow 2Cl^-$ reduction

$\qquad 2I^- + Cl_2 \rightarrow I_2 + 2Cl^-$ overall

(c) $\qquad$ $Cu \rightarrow Cu^{2+} + 2e^-$ oxidation

$\qquad Ag^+ + e^- \rightarrow Ag$ reduction

$\qquad Cu + 2Ag^+ \rightarrow Cu^{2+} + 2Ag$ overall

(d) $\qquad$ $Cu^+ \rightarrow Cu^{2+} + e^-$ oxidation

$\qquad Cu^+ + e^- \rightarrow Cu$ reduction

$\qquad 2Cu^+ \rightarrow Cu^{2+} + Cu$ overall

7. (a) Oxidant: Cl_2, reductant: Fe^{2+}

(b) Oxidant: MnO_2, reductant: HCl

(c) Oxidant: SO_2, reductant: H_2S

(d) Oxidant: N_2, reductant: H_2

8. (b) and (e)

9. (a) 10 800 C

(b) 12 000 C

10. 0.26 mol e^-

11. (a) 0.1 mol e^-

(b) 4 mol e^-

(c) 0.4 mol e^-

(d) 0.5 mol e^-

12. 5.44 g of Mg

13.

    (a) At anode, $2Cl^- \rightarrow Cl_2 + 2e^-$

        at cathode, $Zn^{2+} + 2e^- \rightarrow Zn$

    (b) At anode, $2Br^- \rightarrow Br_2 + 2e^-$

        at cathode, $Na^+ + e^- \rightarrow Na$

    (c) At anode, $2O^{2-} \rightarrow O_2 + 4e^-$

        at cathode, $Al^{3+} + 3e^- \rightarrow Al$

14. (a) 69 120 C

    (b) 0.72 mol e^-

    (c) 0.36 mol of Cu

    (d) 22.78 g of Cu

15.

    (a) $Mg \rightarrow Mg^{2+} + 2e^-$

        $Cl_2 + 2e^- \rightarrow 2Cl^-$

    (b)

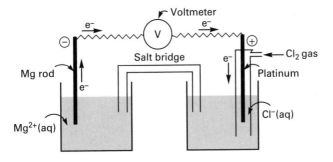

    (c) Standard cell potential = 3.73 V

16. (a) $E^{\ominus}_{cell} = 0.78$ V

    (b) $Fe(s) \mid Fe^{2+}(aq) \parallel Cu^{2+}(aq) \mid Cu(s)$

$$(c) \quad \text{Fe(s)} \rightarrow \text{Fe}^{2+}(aq) + 2e^- \; E^{\ominus} = +0.44 \text{ V}$$

$$\text{Pb}^{2+}(aq) + 2e^- \rightarrow \text{Pb(s)} \; E^{\ominus} = -0.13 \text{ V}$$

$$\text{Fe(s)} + \text{Pb}^{2+}(aq) \rightarrow \text{Fe}^{2+}(aq) + \text{Pb(s)}, \; E^{\ominus}_{cell} = +0.31 \text{ V}$$

Chapter 7

1. $[H^+] = 0.05$ mol dm^{-3}, $[OH^-] = 2.0 \times 10^{-13}$ mol dm^{-3}

2. $[OH^-] = 0.04$ mol dm^{-3}, $[H^+] = 2.5 \times 10^{-13}$ mol dm^{-3}

3. (a) $[H^+] = 2.5$ mol dm^{-3}
    $[OH^-] = 4.0 \times 10^{-15}$ mol dm^{-3}

   (b) $[H^+] = 0.4$ mol dm^{-3}
    $[OH^-] = 2.5 \times 10^{-14}$ mol dm^{-3}

   (c) $[H^+] = 1.0 \times 10^{-3}$ mol dm^{-3}
    $[OH^-] = 1.0 \times 10^{-11}$ mol dm^{-3}

   (d) $[H^+] = 1.0 \times 10^{-11}$ mol dm^{-3}
    $[OH^-] = 1.0 \times 10^{-3}$ mol dm^{-3}

   (e) $[H^+] = 5.0 \times 10^{-14}$ mol dm^{-3}
    $[OH^-] = 0.2$ mol dm^{-3}

   (f) $[H^+] = 1.25 \times 10^{-12}$ mol dm^{-3}
    $[OH^-] = 8.0 \times 10^{-3}$ mol dm^{-3}

4. (a) −4.0   (b) 0.4   (c) 3.0
   (d) 11.0   (e) 13.3   (f) 11.9

5. (a) $[H^+] = 5.0 \times 10^{-6}$ mol dm^{-3}

   (b) $[OH^-] = 2.0 \times 10^{-9}$ mol dm^{-3}

   (c) pOH = 8.7

6. (a) $[H^+] = 0.03$ mol dm^{-3}

   (b) pH = 1.5

7. $[OH^-] = 3.2 \times 10^{-2}$ mol dm^{-3}
    $[H^+] = 3.1 \times 10^{-13}$ mol dm^{-3}

8. Lactic acid > benzoic acid > ethanoic acid >
   hydrocyanic acid > phenol

   $$\xrightarrow{\qquad\qquad\qquad}$$

   acid strength decreases

9. Ethanoic acid < chloroethanoic acid < dichloroethanoic
   acid < trichloroethanoic acid

   $$\xrightarrow{\qquad\qquad\qquad}$$

   acid strength increases

10. (a) CH_3NH_2

    (b) $C_6H_5NH_2$

    (c) $CH_3NH_2(aq) + H_2O(l) \rightleftharpoons CH_3NH_3^+(aq) + OH^-(aq)$

    (d) $K_b = \dfrac{[CH_3NH_3^+]\,[OH^-]}{[CH_3NH_2]}$

    (e) Conjugate acids: NH_4^+, $CH_3NH_3^+$, $C_6H_5NH_3^+$

    $$CH_3NH_3^+ < NH_4^+ < C_6H_5NH_3^+$$

    $$\xrightarrow{\qquad\qquad\qquad}$$

    acid strength increases

11. $K_a = 1.6 \times 10^{-5}$ mol dm^{-3}, p$K_a = 4.8$

12. pH = 11.1

13. pH = 4.5

14. pH = 9.4

Chapter 8

1. C = 48.7%, H = 8.1%, O = 43.2%

2. C = 76.6%, H = 6.4%, O = 17.0%

3. (a) CHO_2

   (b) C_2H_4O

   (c) CH_2O

   (d) $C_{12}H_{22}O_{11}$

4. Empirical formula = CH_2
   Molecular formula = C_6H_{12}

5. Empirical formula = CH,
   Molecular formula = C_6H_6

6. Empirical formula = C_3H_8O,
   Molecular formula = C_3H_8O

7. Empirical formula = CH_2O,
   Molecular formula = $C_6H_{12}O_6$

8. Empirical formula = C_2H_7N,
   Molecular formula = C_2H_7N

9. Empirical formula = C_2H_5Cl,
   Molecular formula = C_2H_5Cl

10. (a) Empirical formula = $C_6H_{12}O$

    (b) Molecular formula = $C_6H_{12}O$

    (c) Assuming that the *m/z* value at 85 is for the fragment
    ion $(M - CH_3)^+$, that at 71 is for $(M - C_2H_5)^+$, and
    that at 43 is for $(CH_3CO)^+$, a probable structural
    formula for the compound is $CH_3CH_2CH_2CH_2COCH_3$.

11. The relative molecular mass of the compound is 72. The
    compound contains an aldehyde group (–CHO) of
    relative mass 29. The difference (72 – 29 = 43) represents
    the mass of the alkyl group attached to the aldehyde
    group.

    $29 = (CHO)^+$ or $(C_2H_5)^+$

    $43 = (M - CHO)^+$ or $(M - C_2H_5)^+$,
    i.e. $(CH_3CH_2CH_2)^+$ or $(CH_2CHO)^+$

    $57 = (M - CH_3)^+$, i.e. $(CH_2CH_2CHO)^+$

    $72 = M^+$, i.e. $(CH_3CH_2CH_2CHO)^+$

    A possible structural formula is $CH_3CH_2CH_2CHO$

12. (a) $74 = M^+$, i.e. $(C_3H_6O_2)^+$, $59 = (M - CH_3)^+$,
    i.e. $(C_2H_3O_2)^+$, $43 = (CH_3CO)^+$,
    $42 = (C_2H_2O)^+$, $28 = (CO)^+$

(b) two CH_3 groups

(c) $1746 \ cm^{-1}$ and $1246 \ cm^{-1}$ for C=O and C–O
    stretching respectively

The structural formula is probably CH_3COOCH_3

13. $CH_3CH_2COOCH_3$

14. $CH_3CH_2COCH_2CH_3$. Pentan-3-one is a symmetrical
    molecule. The four CH_2 protons are equivalent and so
    are the six CH_3 protons.

15. $C_6H_5COOCH_3$. Hydrolysis produces benzoic acid.

16. (a) $120 = (C_8H_8O)^+$, $105 = (M^+ - 15) = (C_7H_5O)^+$,
    $77 = (C_6H_5)^+$, $43 = (CH_3CO)^+$

(b) $C_6H_5COCH_3$

Index